Contents

Introduction

A: Legal and management

- 01 Health and safety law 11
- 02 Construction (Design and Management) Regulations 23
- 03 Developing a health and safety management system 47
- 04 Risk assessments and method statements 59
- 05 Leadership and worker engagement 77
- 06 Statutory inspections, checks and monitoring 93
- 07 Accident reporting and emergency procedures 99

B: Health and welfare

- 08 Health and welfare 117
- 09 Personal protective equipment 131
- 10 Dust and fumes (Respiratory risks) 145
- 11 Noise and vibration 163
- 12 Hazardous substances 175
- 13 Manual handling 187

C: General safety

- 14 Site set up and security ....
- 15 Fire prevention and control ....
- 16 Electrical safety ....
- 17 Work equipment and hand-held tools ....
- 18 Site transport safety 245
- 19 Lifting operations 255

D: High risk activities

- 20 Working at height 269
- 21 Excavations and buried services 297
- 22 Confined space working 311

E: Environment

- 23 Environmental management 323
- 24 Waste management 345

ii

Introduction

Contents

Overview	2
Acknowledgements	2
About the construction industry	3
How to use Site supervision simplified	5
Toolbox talks	6
Use of icons	7
Companion website	7
Augmented reality	8
Further supporting information from CITB	8

Introduction

Overview

Site supervision simplified (GE 706) is aimed at supervisors and first-line managers, but also provides managers and owners of small construction companies with easy-to-understand information and practical guidance to help them implement, supervise and monitor the required standards of health, safety and environment on construction sites.

This is the official reference book for CITB's Site Safety Plus *Site Supervisors' Safety Training Scheme* (SSSTS), a two-day course for first line managers and supervisors, as a supporting document not only on the course but for reference purposes at work afterwards.

It is also a useful source of information on basic, site-based health, safety and environment issues for university undergraduates, college students on construction management courses and for those on other construction-focused courses.

Content provides the supervisor with advice on solutions, case studies and practical day-to-day guidance on what to do, as well as short clear explanations on related legislation.

It is recommended that owners of small and medium-sized companies also refer to *Construction site safety* (GE 700), which covers in more detail the legal duties placed on employers.

Acknowledgements

CITB wishes to acknowledge the assistance offered by the following organisations in the preparation of this edition of GE 706:

- ✓ Countryside Properties plc
- ✓ Easi-Dec Access Systems Ltd
- ✓ eBrit Services Ltd
- ✓ Environment Agency
- ✓ Environmental and Waste Consulting
- ✓ May Gurney Ltd
- ✓ Morgan Sindall
- ✓ National Construction College
- ✓ Skanska.

And a special thank you to:

- ✓ MJ Fuller and Associates.

Introduction

About the construction industry

- Approximately 2.47 million people are employed in the UK construction industry. This includes housing, utilities, repair and maintenance, refurbishment, shopfitting, demolition, roofing, mechanical and electrical, plumbing and highways maintenance.

- The UK construction industry is made up of 197,965 construction businesses and 90% of companies employ less than ten workers.

- Construction workers (just like you) could die due to work-related ill health.

- Work-related respiratory disease covers a range of illnesses that are caused or made worse by breathing in hazardous substances (such as construction dust) that damage the lungs.

- 4,500 people die each year due to past exposure to asbestos.

- 500 people (and more each year) are dying from silica-related lung diseases (dust from cutting blocks, kerbs, and so on). Many more suffer from occupational asthma or are forced to leave the industry due to work-related ill health.

- For 2013 there was an estimated 31,000 new cases of work-related ill health, with rates of musculoskeletal disorder significantly higher than average.

- On average 50 workers are killed each year due to accidents. The biggest killer (around half) is due to falls from height.

- Each year approximately 2,500 workers are seriously injured (for example, broken bones, fractured skull, amputations) and 5,700 have reportable injuries.

- The most common causes of reported injuries are due to manual handling and slips, trips and falls and being struck by moving or falling objects. 60% of all work at height injuries are due to falls from below head height.

- Fatal injuries to members of the public are steadily falling. Just over a quarter (27%) of fatal injuries to the public over the past five years were due to falls, with slips and trips accounting for 18% and those involving moving vehicles accounting for 14%.

Why so many accidents?

Many reports of present day construction accidents and ill health make depressing reading because simple actions were not taken to prevent them. In many cases, those planning the jobs totally failed to consider the health or safety of the people carrying out the work (and possibly others who were affected) and actively manage the situation.

Common examples of such events include:

- the increasing number of workers who suffer from cancers and life-changing illnesses from breathing to skin complaints – some of these force the sufferer to give up work, because exposure to dangerous substances, such as dust, is not even considered, let alone prevented or controlled

- the deaths and serious injuries that occur because people fall from height – often basic actions (like using temporary work platforms on fragile roofs, installing edge protection or using a safety harness and lanyard clipped to a strong point) were simply not taken

- workers being buried in collapsed excavations because the sides were unstable or not supported

- workers being killed or injured by construction plant because pedestrians were not kept out of the plant operating area.

Achieving a reasonable standard of on-site health and safety is not difficult. Where the work to be carried out is relatively uncomplicated and familiar, the precautions that need to be taken are in many cases simple and commonsense or may require a little investigation or

Introduction

reading. The crucial decision for anyone with responsibility for on-site health and safety is to know when they have reached the limits of their knowledge and capabilities and therefore need the assistance of someone with specialist knowledge.

Caution should also be exercised when a job is not going to plan and there is the temptation to resort to improvised methods of working. If the person in control is not at ease with the way that things are going they should stop the job, step back and think things through carefully before deciding upon a course of action.

Research from the Health and Safety Executive (HSE) has shown that workers are most vulnerable during their first few days on site.

Setting out

Construction is an exciting industry. It is constantly changing as projects move on and jobs get done. As a result of this a building site is one of the most dangerous environments to work in.

But many accidents that occur on sites can be avoided if everyone on site works together. So a free film *Setting out*, produced by the industry, sets out what the site must do and what you must do to stay healthy and safe at work.

This film is essential viewing for everyone involved in construction, and should be viewed before sitting the CITB *Health, safety and environment test*. The content of the film is captured in summary here, and these principles form the basis for the behavioural case study questions, which became a new element of the test in Spring 2012.

For further information or to view *Setting out* refer to the CITB website at citb.co.uk

Part 1. What you should expect from the construction industry

Your site and your employer should be doing all they can to keep you and your workforce safe.

Before any work begins the site management team will have been planning and preparing the site for your arrival. It is their job to ensure that you can do your job safely and efficiently.

Five things **the site** you are working on **must** do:

- ✓ know when you are on site (signing in and out)
- ✓ give you a site induction
- ✓ give you site-specific information
- ✓ encourage communication
- ✓ keep you up-to-date and informed.

Part 2. What the industry expects of you

Once the work begins, it is up to every individual to take responsibility for carrying out the plan safely. This means you should follow the rules and guidelines as well as being alert to the continuing changes on site.

Five things **you must** do:

- ✓ respect and follow the site rules
- ✓ safely prepare each task
- ✓ do each task responsibly
- ✓ know when to stop (if you think anything is unsafe)
- ✓ keep learning.

Every day the work we do improves the world around us. It is time for us to work together to build an industry that puts its people first. By working together we can build a better industry that respects those who work in it.

Introduction

Working Well Together

The Working Well Together (WWT) campaign is an industry-led initiative that helps support micro and small businesses improve their health and safety performance. The campaign undertakes a variety of activities, including health and safety awareness days, designer awareness days, breakfast and evening events, roadshows and regional WWT groups.

 To find out how the WWT campaign can help you and your company refer to www.wwt.uk.com

Poster from the Working Well Together campaign

Continuing to expand your health and safety knowledge and competence

After reading this book or attending the CITB Site Safety Plus two day Site Supervisors' Safety Training Scheme you may wish to consider the next step in expanding your health and safety awareness and competence.

In due course, if you already are, or are about to become responsible for planning, organising, monitoring, controlling and administering a workforce, you could consider a more in-depth look at health and safety by attending the five day CITB Site Management Safety Training scheme.

 Details of course training providers can be found at www.citb.co.uk

How to use Site supervision simplified

GE 706 follows the standard structure that is used across all core CITB publications:

Section A: Legal and management
Section B: Health and welfare
Section C: General safety
Section D: High risk activities
Section E: Environment

Each chapter begins with a summary list of the employer's responsibilities, together with a corresponding checklist for supervisors, to help them understand what they and their employer should be doing together, to protect their workforce.

Introduction

What's new in the 2014 edition?

Minor amendments that reflect changes in legislation include:

- ☑ RIDDOR reporting of accidents and dangerous occurrences
- ☑ Health and Safety (First Aid at Work) Regulations
- ☑ Construction (Head Protection) Regulations *(withdrawn)*
- ☑ Notification of Conventional Tower Cranes Regulations *(withdrawn)*.

There is also a new chapter on site set up and security (C14) and an expanded section on fragile roofs in Chapter D20.

Within sections A to D the chapter order has been amended to more closely align with chapters in *Construction site safety (GE 700)*.

As well as the inclusion of new images and minor text changes, cross referencing has been updated to provide links to GE 700.

Toolbox talks

Toolbox talks (GT 700) has been updated to follow the same structure and chapter layout as *Site supervision simplified* (GE 706) in order to help supervisors prepare for a toolbox talk.

For example, if a supervisor wants to deliver a talk on concrete and silica, they can first look up the relevant GT 700 toolbox talk: B14 Silica dust, which is in Section B, under the Dust and fumes (Respiratory risks) topic.

As each topic relates to a chapter within GE 706, further information to help prepare can easily be found within GE 706, Section B Health and welfare, 10 Dust and fumes (Respiratory risks).

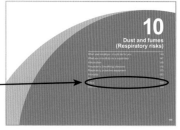

GE 706 is the official reference material for CITB Site Safety Plus *Site Supervisors' Safety Training Scheme* (SSSTS), a two-day course for first line managers and supervisors.

Introduction

Use of icons

A set of icons emphasises key points within the text and also directs readers to further information. The icons are explained below.

	Website/further info		Caution
e.g.	Example		Consultation
?	Question		Guidance
	Ideas		Case study
	Notes		Quote
★	Favourite		Definition
!	Important		Checklist
	Good practice		Video
	Poor practice		

Companion website

A companion website has been created to support *Construction site safety* (GE 700). This contains up-to-date information, which would also be of use to supervisors and users of GE 706, including:

- ✓ news (such as legislation changes, industry guidance and good practice)
- ✓ any minor amendments or updates to the current edition of GE 700
- ✓ web-links, phone numbers and addresses.

The companion website can be found at www.citb.co.uk/GE700companion

Save the companion website address to your favourites, so it is always available when you need it.

Introduction

Augmented reality

How to install the app

- ☑ Go to the appropriate app store (Apple or Android) and download the Layar app (free of charge) to your mobile phone or tablet.
- ☑ Look out for this logo [right], which is on the cover of this book and indicates that you are on a Layar-friendly page.
- ☑ Open the Layar app and scan the Layar-friendly page.
- ☑ Wait for the page to activate on your device.
- ☑ Touch one of the buttons that have appeared to access additional content.

Further supporting information from CITB

CITB has a wide range of products, publications and courses that could help to improve your health, safety and environment knowledge.

To discover more about CITB and the services, publications and courses offered visit www.citb.co.uk

Health, safety and environment publications

Audience and publication	Format
Site managers	
GE 700 *Construction site safety*	Printed, CD and online
RACD *Risk assessment and method statement manager*	CD
SA 03 CD *Construction site health, safety and environment auditing system*	CD
Supervisors	
GE 706 *Site supervision simplified*	Printed, CD and online
GT 700 *Toolbox talks*	Printed, CD and online
GT 701 *Safety critical communication toolbox talks*	Printed
Operatives	
GE 707 *Safe start*	Printed
Health safety and environment test delegates	
GT 100 *Health, safety and environment test for operatives and specialists*	Printed, DVD, download and app
GT 200 *Health, safety and environment test for managers and professionals*	Printed, DVD, download and app

Introduction

Site Safety Plus courses

Audience/topic and course	Duration
Directors	
Directors' Role For Health and Safety Course	One day
Site managers	
Site Management Safety Training Scheme (SMSTS)	Five days
Plant managers	
Plant Management Safety Training Scheme (PMSTS)	Five days
Supervisors	
Site Supervisors' Safety Training Scheme (SSSTS)	Two days
Temporary works co-ordinators	
Temporary Works Co-ordinator Course	Two days
Operatives	
Health, Safety and Environmental Awareness	One day
Environment	
Site Environmental Awareness Training Scheme (SEATS)	One day
Behavioural safety	
Achieving Behavioural Change (ABC)	One day
Shopfitting	
Site Safety for Shopfitters and Interior Contractors	Three days

National Construction College

The National Construction College is focused on creating a highly skilled, safe and professional UK construction workforce. To achieve this, it has more first-class instructors in more locations than any other construction training provider in Europe and offers free professional advice on finding the right training for individuals and companies.

Introduction

01

Health and safety law

What your employer should do for you	12
What you should do as a supervisor	13
Introduction	14
The progress of health and safety law	14
Health and Safety at Work Act 1974	15
Approved Codes of Practice	15
Health and safety regulations	16
Standards of compliance	17
Legal duties	18
Corporate Manslaughter and Corporate Homicide Act	19
Health and Safety (Offences) Act	20
Enforcement	20
Supporting information	22

Health and safety law

What your employer should do for you
1. Understand the duties imposed by the Health and Safety at Work Act 1974.
2. Understand and put into practice the standard of compliance required.
3. Be aware of the implications of the duties imposed by the Corporate Manslaughter and Corporate Homicide Act.
4. Understand the legal status of health and safety regulations.
5. Recognise the importance of the Construction (Design and Management) Regulations.
6. Make employees aware of their statutory duties and ensure understanding and compliance.
7. Understand the functions and powers of the Health and Safety Executive (HSE).
8. Ensure access to competent health and safety advice.

What you should do as a supervisor

Checklist		Yes	No	N/A
1.	Put into practice your duties imposed by the company health and safety policy.			
2.	Understand the standard of compliance required to ensure safety of your work team and others.			
3.	Comply with your safety instructions and ensure your work team understand their duties and responsibilities.			
4.	Supervise your work team to ensure their, and your safety.			
5.	Ensure that work equipment is in good condition, inspected and used correctly.			
6.	Ensure correct personal protective equipment (PPE) is provided, correctly worn/used and replaced when defective.			
7.	Co-operate with your employer so that they can carry out their legal duties.			
8.	Report anything that you feel is unsafe.			

A 01

Introduction

The purpose of health and safety legislation is to protect the wellbeing of people at work and others (such as the general public) by ensuring that work is carried out in a manner that is safe and free of risks to health for everyone who may be affected.

There have always been accidents and ill health resulting from people carrying out work activities. Fortunately we have come a long way from earlier days when such events were simply accepted as a part of having a job – many people went to work and took their chances. Even so, workplace accidents and work-related ill health still occur, with the building and construction industry having a disproportionately high rate of both.

Everyone who goes to work has a right to return home uninjured and in a good state of health. Similarly, visitors to sites and passers-by have the same right to the protection of their health and safety.

When health and safety law is broken, a criminal offence is committed. If the results are sufficiently serious the offence could be punishable by a fine or even imprisonment. The Health and Safety Executive (HSE) can also issue enforcement notices or offer advice. In accordance with regulations, the HSE may charge for its time when investigating a breach of health and safety law. This is known as fee for intervention (FFI).

Whilst health and safety legislation is embedded in criminal law, in some circumstances legislation gives a person who has been injured or made ill through work the right to take legal action against the employer for compensation, through the civil courts.

 You can insure against civil liabilities but you cannot insure against criminal liabilities.

The primary focus of health and safety legislation is to put legal duties on employers (and the self-employed, who in many cases have the same legal duties as employers) to ensure that work is carried out safely and without risks to health.

It should be noted that some legislation places legal duties upon employees. In specific situations, legal duties are also placed upon *duty holders* as defined in the relevant legislation. The actions that must be taken to fulfill these duties are explained in the appropriate chapters of this book.

In many situations, site-based staff (such as supervisors, site managers or project managers) will be nominated by the employer to ensure that the employer's legal duties are complied with at site level. However, legal duties still lie with the employer who must be confident in the ability and competence of their supervisors, managers and others to manage health and safety on their behalf.

The progress of health and safety law

- ☑ The first Act of Parliament directly concerned with safety was the Factories Act passed in 1802, which dealt with the morals of apprentices.

- ☑ From 1812 onwards, there was a succession of Acts to regulate working conditions in factories, with particular reference to women and children.

- ☑ In 1833, the first four factory inspectors were appointed. Children aged nine were being sent up chimneys as chimney sweeps, where they could work a 48-hour week. They were supposed to spend two hours a day at school. People as young as 13 sometimes worked a 69-hour week.

- ☑ The first two female factory inspectors were appointed in 1893, the first medical inspector in 1898.

Health and safety law

- In 1901, a comprehensive Factories and Workshops Act was passed. It lasted until the Factories Act of 1937 replaced it, then this, in turn, was repealed by the Factories Act of 1961.

- The Health and Safety at Work Act, the basis for modern legislation, came into force in 1974, bringing with it protection for virtually everyone at work and consolidating much of the earlier industry-specific legislation under a single Act.

- The Construction (Design and Management) Regulations were originally drafted to address the continuing unacceptably high rate of accidents and ill health befalling construction workers. However, they brought about a culture of bureaucracy and form-filling rather than a focus on risk management.

- The current Construction (Design and Management) Regulations (CDM) are the cornerstone legislation for the construction industry and clearly reinforce the requirements for competence, co-operation and co-ordination, with a focus on 'the right information to the right people at the right time'.

- The coming into force of the Corporate Manslaughter and Corporate Homicide Act and the Health and Safety (Offences) Act, demonstrated a hardening of Government's attitude with regard to serious breaches of health and safety legislation.

- Professor Ragnar Löfstedt's report, *Reclaiming health and safety for all: an independent review of health and safety legislation,* was published in 2011. It was commissioned as part of the Government's plan to overhaul the health and safety system in Britain. The report considers ways in which health and safety legislation can be combined, simplified or reduced so that the burden on British businesses can be alleviated. It has led to a review of a number of regulations, including CDM.

Health and Safety at Work Act 1974

The Health and Safety at Work Act is the primary piece of health and safety legislation in the United Kingdom. The Act is termed as an enabling Act, which allows the Government to make health and safety regulations that become part of the law. The requirements of the Act are very general and wide-ranging, such as:

> It shall be the duty of every employer to ensure, so far as is reasonably practicable, the health, safety and welfare at work of all their employees.

This typical requirement does not specify any technical requirements or set any minimum standards of behaviour that can be measured, but it does clearly outline the requirement for safe places of work.

Health and safety regulations expand upon the legal requirements of the Act with regard to specific occupational activities and include specific and technical requirements. Health and safety regulations are often supported by guidance notes or Approved Codes of Practice (ACoPs), which explain in plain language how compliance with the law can be achieved.

Approved Codes of Practice

The Health and Safety at Work Act makes provision for the production of ACoPs, where appropriate, in support of some health and safety regulations.

Health and safety law

Health and safety regulations

Many sets of regulations have been developed and introduced as a means of incorporating European Community legislation into our domestic legal framework. There are many sets of health and safety regulations in existence that are relevant to the work activities carried out by the construction industry. Some sets of regulations apply to construction activities only.

Health and safety regulations are a part of UK law that place duties on employers and employees. It is a criminal offence to contravene them.

Examples of legislation that have an effect on the building and construction industry are shown below.

Management of Health and Safety at Work Regulations

The main requirement on employers is to carry out a risk assessment and employers with five or more employees need to record the significant findings of the risk assessment.

Construction (Design and Management) Regulations *(refer to Chapter A02)*

There is a strong focus on the idea of competence, which is emphasised in the Approved Code of Practice (ACoP). To be competent, an organisation or individual must:

- ☑ have knowledge of the specific tasks to be undertaken and the risks that the work will entail
- ☑ have sufficient experience and ability to carry out their duties in relation to the project
- ☑ recognise their limitations
- ☑ take appropriate action in order to prevent harm to those carrying out construction work, or those affected by the work.

Work at Height Regulations *(refer to Chapter D20)*

Duty holders are required to ensure that:

- ☑ all work at height is properly planned and organised
- ☑ all work at height takes account of weather conditions that could endanger health and safety
- ☑ those involved in work at height are trained and competent
- ☑ the place where work at height is carried out is safe
- ☑ equipment for work at height is appropriately inspected
- ☑ the risks from fragile surfaces are properly controlled, as are the risks from falling objects.

Provision and Use of Work Equipment Regulations *(refer to Chapter C17)*

Correct equipment must be provided, inspected, tested and properly maintained.

Electricity at Work Regulations *(refer to Chapter C16)*

People in control of electrical systems must ensure that they are safe to use and they are maintained in a safe condition.

Control of Asbestos Regulations *(refer to Chapter B10)*

Requires the identification of asbestos-containing materials (ACMs) that may be present, before any work is carried out.

Reporting of Injuries, Diseases and Dangerous Occurrences Regulations (RIDDOR) *(refer to Chapter A07)*

A legal duty is placed on employers, the self-employed and those in control of premises to report:

- ☑ work-related deaths or specified injuries
- ☑ injuries resulting in an absence of more than seven days

Health and safety law

- ✓ work-related diseases
- ✓ dangerous occurrences.

Standards of compliance

Within health and safety legislation specific words or phases are used to qualify or describe the standard of compliance that must be achieved with regard to some legal duties. The meanings of these are explained below.

It shall be the duty of an employer to…

This means that the employer *must* comply with the legal duty being described. The word 'shall' (if not qualified by the phrase 'so far as reasonably practicable' or the word 'practicable') leaves no scope for not complying with the duty.

It shall be the duty of an employer to… as far as is reasonably practicable

Where a requirement to carry out a specific legal duty is qualified by the phrase 'as far as is reasonably practicable', employers are allowed to exercise their judgement on the extent of the measures that need to be taken to ensure the health and safety of the person(s) carrying out the job and anyone else who may be affected by it.

Deciding what are reasonably practicable measures to take should be based upon the findings of a risk assessment.

 Reasonably practicable means that the risks involved in carrying out the work may be balanced against the cost in terms of money, inconvenience and time.

Where the risks to health and safety in carrying out a job are found to be low in comparison to what would be disproportionately high costs to overcome the risks totally, the employer need only take the measures that are considered to be reasonably practicable.

For example – Reasonably practicable

It was necessary to provide access to working platforms of a scaffold at four different levels, the higher level being a narrow lift between two structures for a 12-week period. A decision had to be made on the best means of access.

Legislation on working at height requires 'every employer to ensure that work at height is carried out in a manner which is, so far as reasonably practicable, safe'.

In planning the job, the risk assessment showed:

- ✓ it would be necessary for tools and materials for various trades to be carried up and down from the main working platform but not from the higher level lift
- ✓ the ground was firm and level and there was plenty of space at the bottom of the scaffold
- ✓ if a ladder was used, it could be securely tied to the scaffold and suitable handholds could be provided at the stepping-off point
- ✓ the tools and materials would have to be hoisted up using a small electric hoist fixed to the scaffold
- ✓ various trades could be working together and at the same time
- ✓ only a means of access was required, not a place of work, to the higher level lift
- ✓ there were no weather considerations that would make the use of a ladder unduly unsafe.

Health and safety law

Given the circumstances, it was decided that a ladder was not a reasonably practicable measure to prevent a fall with regard to access to the main working platform where tools and materials were required and therefore a stair tower was used. For access to the higher level lift a ladder was considered to be suitable, complete with a ladder safety gate. The stair tower also offered quicker, safer and easier access.

 Practicable means that there is no scope for taking cost and convenience into account; the duty must be complied with if it is capable of being carried out within the current state of knowledge and technology (if it is technically possible there is no choice).

For example – Practicable

It was necessary for someone to use a disc-cutter to cut paving slabs to the required size.

The legal duty requires that the rotating blade of the machine be guarded to the extent that it is practicable to do so. This acknowledges that total guarding of the blade is not possible because the machine could not then be used for the job for which it was designed.

The law requires that the guard be adjusted to expose enough blade to enable the job to be carried out safely whilst providing the maximum degree of protection for the operator.

Not to use a guard at all is not an option, even if it causes cost or inconvenience.

Burden of proof

Many of the duties on employers are qualified with the phrases 'so far as is reasonably practicable' or 'so far as is practicable'. Generally, in a court of law a defendant is innocent until proven guilty. However, with regard to these phrases, the burden of proof is reversed and so the defendant must prove that it was not reasonably practicable, or practicable, as the case may be, to do more than was in fact done to mitigate the risks. In effect the defendant is guilty until they prove their innocence.

Legal duties (general)

The broad legal duties of employers and employees, as specified in the Health and Safety at Work Act, are shown below.

Employers' responsibilities

An employer must, so far as is reasonably practicable:

- ☑ protect the health, safety and welfare at work of all their employees

- ☑ provide and maintain plant and systems of work that are safe and without risk to health

- ☑ have arrangements for ensuring safety and absence of risk to health in connection with the use, handling, storage and transport of articles and substances

- ☑ provide such information, instruction, training and supervision as is necessary to ensure the health and safety at work of employees

- ☑ maintain any place of work under their control in a condition that is safe and without risks to health, and with access to and egress from it, that are safe and without such risks

Health and safety law

- provide and maintain a working environment that is safe, without risks to health and adequate as regards the welfare of employees.

Employees' duties

It is the duty of every employee:

- to take reasonable care for the health and safety of themselves or others who may be affected by their acts or omissions
- to co-operate with the employer in all matters relating to health and safety
- not to intentionally or recklessly interfere with or misuse anything provided in the interests of health, safety and welfare
- to use anything provided by the employer in accordance with instructions
- to report anything that is thought to be dangerous.

The requirement, on the employer, to do what is reasonably practicable to ensure the health, safety and welfare of employees at work is balanced by a requirement, on the employees, to comply with any necessary rules or instructions and to take reasonable care of themselves or others.

Legal duties (specific)

The above general legal duties are expanded upon and made more specific by the various sets of regulations that are relevant to construction industry activities.

These more specific duties will be explained at the appropriate places in this book.

 Most chapters of this book outline legal duties placed on employers. Readers should note that these duties apply equally to self-employed persons.

Corporate Manslaughter and Corporate Homicide Act

Since 6 April 2008, companies whose gross negligence leads to the death of individuals can face prosecution for manslaughter (homicide in Scotland) under the Corporate Manslaughter and Corporate Homicide Act.

Under this legislation, companies, organisations and, for the first time, Government bodies face an unlimited fine if they are found to have caused death due to their gross corporate health and safety failures. The legislation primarily came about as a result of the failure to identify 'the controlling mind' in companies with complex management structures during court cases, which followed several high profile disasters.

40 m-high scaffolding collapse resulting in one fatality and two other people seriously injured

Health and safety law

However, presentations by authoritative legal professionals have indicated that the legislation might have much wider ranging implications (such as prosecutions following at-work-related road deaths and fatalities).

The Act offers employees of companies, consumers and other individuals greater protection against corporate negligence. It will also focus the minds of those in companies and other organisations by ensuring that they take their health and safety obligations seriously.

The Corporate Manslaughter Act:
- ☑ makes it easier to prosecute companies and other large organisations when gross failures in the management of health and safety lead to death by delivering a new, more effective basis for corporate liability
- ☑ means that both small and large companies can be held liable for manslaughter where gross failures in the management of health and safety cause death, not just health and safety violations
- ☑ complements the current law under which individuals can be prosecuted for gross negligence manslaughter and health and safety offences, where there is direct evidence of their culpability.

Note 1. The Act uses the term 'senior management' and defines it as meaning those persons who play a significant role in the management of the whole or a substantial part of the organisation's activities. This covers both those in the direct chain of management as well as those in, for example, strategic or regulatory compliance roles.

Note 2. It has been speculated that a significant number of future cases brought under the Corporate Manslaughter and Corporate Homicide Act will focus on how employers manage occupational road risk, both for professional drivers and other employees who are required to drive their own or company vehicles in the course of their work. Employers must be aware that under such circumstances vehicles are regarded as a place of work. Consideration should be to proactively manage occupational road risk, for example:

- ☑ *checking driver qualification compliance (professional drivers)*
- ☑ *providing follow-up assessments or advanced driver training*
- ☑ *mobile phone use policy*
- ☑ *checks of vehicle servicing records, insurance and MOT paperwork and so on*
- ☑ *at-work driver policies (maximum driving hours without a break and so on).*

Health and Safety (Offences) Act

This Act came into force in January 2009. The main provisions of the Act are to raise the maximum penalty that can be imposed by a Magistrates' Court for a breach of health and safety regulations from £5,000 to £20,000. The range of offences for which an individual can be imprisoned has also been extended.

The HSE has made it clear that their enforcement policy targets those who cut corners to gain a financial advantage over competitors, by failing to comply with health and safety law and therefore putting workers and others at risk.

Enforcement

Health and safety law

The enforcement of all health and safety law is carried out under the provisions of the Act. This means that methods of enforcement are the same throughout the country.

The Health and Safety Executive (HSE) is responsible for maintaining a force of health and safety inspectors, some of whom specialise in construction industry activities.

Health and safety law

Powers of HSE inspectors

A health and safety inspector:

- ☑ may enter any premises at any reasonable time, or at any time if they suspect a dangerous situation
- ☑ can examine and investigate as necessary
- ☑ can measure, take photographs, samples or possession of anything, if required for evidence
- ☑ can inspect books and other documents
- ☑ can insist that dangerous equipment is made safe
- ☑ can question people and require that they make a signed declaration of the truth of answers given
- ☑ can demand that the scene of an accident remains undisturbed
- ☑ can call upon police to assist entry into premises if necessary

 Health and safety inspectors can also serve enforcement notices on an employer or instigate a prosecution.

If it appears to a HSE inspector that there may be a contravention or breach of health and safety law, they may serve an **improvement notice** on an employer, requiring that improvements to the way health and safety are being managed are completed within a specified time.

If the situation appears to be too dangerous to allow work to continue, the HSE inspector may serve a **prohibition notice**, which immediately stops the work operation stated in the notice until remedial work is satisfactorily completed.

Where serious breaches of health and safety law have occurred, particularly where personal injury or death has resulted from unsafe working practices, the inspector conducting the investigation can instigate the prosecution of the alleged offender(s).

Employers have a right of appeal against any improvement or prohibition notices, which they must lodge within 21 days. When an appeal is lodged against an improvement notice, the requirements of the notice are suspended until the appeal is heard.

However, if an appeal is lodged against a prohibition notice, the effects of the notice remain in force.

Whilst the HSE has the power to visit workplaces and take enforcement action where appropriate, it would rather offer advice and guidance to employers to prevent dangerous situations from occurring.

In October 2012 a fee for intervention (FFI) cost recovery scheme was introduced that applies to businesses and individuals regulated by the HSE. FFI enables the HSE to recover costs incurred when carrying out its duties, where there is a material breach of health and safety law.

HSE inspectors can issue notices if they think it necessary

Environmental law

Environmental law is enforced by the Environment Agency (EA), Natural Resources Wales and Scottish Environment Protection Agency (SEPA).

For further information refer to
Chapter E23 Environmental management and
Chapter E24 Waste management.

Supporting information

The HSE offers a range of services to assist employers and employees. Many publications are produced each year – with some at no charge or downloadable from its website.

For further information refer to the HSE's website at
www.hse.gov.uk

The Environment Agency in England offers information and advice to assist companies in protecting and improving the environment and contributing to sustainable development.

For further information refer to the Environment
Agency website at www.environment-agency.gov.uk

02

Construction (Design and Management) Regulations

What your employer should do for you	24
What you should do as a supervisor	25
Introduction	26
Structure of the regulations	28
Duty holders	29
Project documents	33
Pre-construction information	34
The construction phase plan	37
The health and safety file	41
Consultation with the workforce	42
Competence	42
Co-operation and co-ordination	43
Health and safety on construction sites	44

Construction (Design and Management) Regulations

A 02

	What your employer should do for you
1.	Explain Construction (Design and Management) Regulations (CDM) duties to their employees and how they apply on projects.
2.	Act as a principal contractor to develop the construction phase plan and explain how to implement it.
3.	If acting as a main contractor on non-notifiable projects, provide relevant and understandable information.
4.	Explain the difference and practical implications between notifiable and non-notifiable projects.
5.	As a contractor, be aware of and explain company obligations to the principal contractor and provide information.
6.	Provide and explain relevant paperwork (method statements and risk assessments) to managers and supervisors.
7.	Ensure designs for work (for a client or principal contractor) have removed all reasonably practicable hazards.
8.	Assess the competence of all the work team and provide adequate training and supervision.

The Construction (Design and Management) Regulations

What you should do as a supervisor

Checklist	Yes	No	N/A
1. Put into practice safe working methods to ensure a safe place of work for your work team.			
2. If a principal contractor, to assist in implementing the construction phase plan and managing contractors.			
3. If acting as the main contractor, to understand that you have a duty to ensure compliance of your work team.			
4. Provide supervision and advice, taking into account the abilities of your work team.			
5. Ensure that inspections and tests are carried out and recorded.			
6. Consult with your work team and advise your manager of items requiring attention.			
7. Ensure that any designs that require modification are checked before the work is carried out and recorded for the future.			
8. Ensure the welfare facilities are in a suitable state of cleanliness with adequate provision to match site needs.			

A 02

A 02 Construction (Design and Management) Regulations

Introduction

From a health, safety and welfare standpoint, the Construction (Design and Management) Regulations (CDM) are the cornerstone of construction design and management. These regulations came into force on 6 April 2007 and replace both the earlier Construction (Design and Management) Regulations and the Construction (Health, Safety and Welfare) Regulations. They also amend the Workplace (Health, Safety and Welfare) Regulations with the effect that some of those regulations also now apply to construction sites.

The Health and Safety Executive (HSE) is currently reviewing the CDM Regulations, and they are likely to change in 2014-15.

The regulations are supported by:

- ☑ an Approved Code of Practice (ACoP), *Managing health and safety in construction* (Code No. L144), which expands upon what is expected of duty holders to comply with the regulations; L144 is available from HSE books, freely downloadable at www.hse.gov.uk/construction/cdm.htm

- ☑ guidance notes for each duty holder, which were written by the industry and are available at www.citb.co.uk/Health-Safety-and-other-topics/Health-Safety/health-safety-legislation

The Construction (Design and Management) Regulations

The Construction (Design and Management) Regulations, abbreviated to CDM for clarity and simplicity in this chapter, introduce some significant changes. These include:

- ✓ a greater emphasis on the duties of construction clients
- ✓ removing the facility for the client to transfer their criminal liability to a client's agent
- ✓ replacing the planning supervisor with a CDM co-ordinator
- ✓ simplifying the criteria by which projects are notifiable to the HSE
- ✓ placing a greater emphasis on competence of all duty holders and increasing the duties on clients and designers
- ✓ giving greater consideration to designing the safe use of the building
- ✓ removing the requirement to notify the HSE of any projects carried out for a domestic client.

The overall concept is to improve health and safety standards on site without creating unnecessary paperwork and bureaucracy.

The right information to the right people at the right time is of key importance.

For further information refer to GE 700 *Construction site safety*, **Chapter A03 Construction (Design and Management) Regulations.**

When the regulations apply

General health, safety and welfare duties previously required under the Construction (Health, Safety and Welfare) Regulations are now contained under CDM so, in effect: every construction project or building job, no matter how small, is a 'CDM job'. Some have further obligations depending upon duration or person days.

The following parts of the regulations apply to **all** projects.

- ✓ Part 1: Introduction.
- ✓ Part 2: General management duties.
- ✓ Part 4: Duties relating to health and safety on construction sites.
- ✓ Part 5: General.
- ✓ Schedule 2: Welfare facilities *(refer to Chapter B08 Health and welfare)*.
- ✓ Schedule 3: Particulars to be included in a report of inspection.

The following also apply to those projects that are notifiable to the HSE *(see next page)*.

- ✓ Part 3: Additional duties where the project is notifiable.
- ✓ Schedule 1: Particulars to be notified to the HSE (Form 10).

Notification by the CDM co-ordinator

The regulations require that details of notifiable projects are sent at the earliest opportunity, preferably online, to the HSE on a Form F10.

A blank form can be found at www.hse.gov.uk/forms

Construction (Design and Management) Regulations

Projects are notifiable if the construction phase is not for a domestic client and:

- ☑ will last longer than 30 working days, or
- ☑ will involve more than 500 person days of construction work.

> Domestic clients are private householders who engage a contractor or builder to carry out work on their private house providing the house is not used in connection with a business. Where a job is carried out on a domestic property as part of, for example, a property development project or on behalf of a body (such as a housing association), the work is not actually for a domestic client and the regulations will apply as appropriate for the work being carried out.

Where a project is notifiable:

- ☑ a CDM co-ordinator and principal contractor must be appointed by the client
- ☑ additional duties are placed upon the client, designers and contractors
- ☑ a construction phase plan and a health and safety file must be compiled
- ☑ there is an emphasis on the importance of liaison between duty holders and on consultation with the workforce.

If not part of a notifiable project, high-risk construction activities will still need to be well planned and managed with a written plan of work (that includes, for example, risk assessments, method statements and possibly supported by a permit to work).

Examples of high risk activities include working in deep excavations or confined spaces; demolition; heavy or complex lifting operations; working with asbestos; and work on or near to exposed, live electrical conductors. The level of detail should be proportionate to the risks involved.

CDM promotes teamwork

Structure of the regulations

The regulations are structured to ensure that:

- ☑ there is a co-ordinated approach to health and safety on site, particularly where there are several contractors on site at any one time
- ☑ adequate time and resources are committed at a sufficiently early stage to draw health and safety into the design and planning phases

The Construction (Design and Management) Regulations

- ☑ there is adequate co-operation and communication between everyone on site who has responsibility for health and safety
- ☑ everyone on site is trained as may be necessary and is competent to do their job
- ☑ health and safety issues are considered at the design and planning stage for the whole life cycle of a new structure, which includes:
  - − construction
  - − maintenance
  - − alteration or extension
  - − everyday use
  - − cleaning
  - − demolition.

Duty holders

Under the regulations, the following have specific legal duties:

- ☑ clients
- ☑ designers
- ☑ CDM co-ordinator
- ☑ principal contractor
- ☑ contractors, including the self-employed
- ☑ anyone who controls construction work.

Whilst duties are not placed specifically on employees by these regulations, duties are placed on every person, which must be taken to include employees.

Duty holders need to be clear about their responsibilities

Client duties

Client duties under CDM are summarised as follows.

- ☑ Where the client fails to appoint a CDM co-ordinator or a principal contractor (or both), the client will assume their legal duties until the appointments are made.
- ☑ Clients may request assistance from the CDM co-ordinator to ensure compliance with the regulations with regard to their duties.

29

Construction (Design and Management) Regulations

A 02

Client duties	
All construction work (Part 2 of the regulations)	Additional duties for notifiable projects (Part 3 of the regulations)
Check competence and resources of all appointees. Ensure there are suitable management arrangements for the project, including welfare facilities. Allow sufficient time and resources for all stages. Provide pre-construction information to designers and contractors.	Appoint a CDM co-ordinator. Appoint a principal contractor. There must be a CDM co-ordinator and principal contractor until the end of the construction phase. Make sure that the construction phase does not start unless there: ■ are suitable welfare facilities ■ is a construction phase plan. Provide information relating to the health and safety file to the CDM co-ordinator. Retain and provide access to the health and safety file.

Principal contractor duties	
All construction work (Part 2 of the regulations)	Additional duties for notifiable projects (Part 3 of the regulations)
	Plan, manage and monitor the construction phase in liaison with contractors. Prepare, develop and implement a written plan and site rules. (An initial plan must be completed before the construction phase begins.) Give contractors the relevant parts of the plan. Make sure suitable welfare facilities are provided from the start and are maintained throughout the construction phase. Check the competence of all their appointees. Ensure that all workers have site inductions and any further information and training needed to carry out the work. Consult with the workers. Liaise with the CDM co-ordinator regarding ongoing design. Secure the site.

Principal contractor duties

The principal contractor must be a contractor in their own right and may change during the course of the project. However, irrespective of who is fulfilling the role, the following requirements are imposed upon them.

The Construction (Design and Management) Regulations

Contractor duties

Contractors are defined as everyone working on site, including the self-employed, but excluding the principal contractor.

The duties that are placed upon contractors by these regulations reflect the good health and safety principles and practices that are a feature of any well-managed site.

Contractor duties

All construction work (Part 2 of the regulations)	Additional duties for notifiable projects (Part 3 of the regulations)
Plan, manage and monitor own work and that of their workers.	Check that the client is aware of their duties and that a CDM co-ordinator has been appointed and the HSE notified before starting work.
Check the competence of all their appointees and workers.	Co-operate with the principal contractor in planning and managing work, including reasonable directions and site rules.
Train their own employees.	
Provide information to their workers.	
Comply with the specific requirements in Part 4 of the regulations.	Provide the principal contractor with details of any contractor(s) engaged.
Ensure there are adequate welfare facilities for their workers.	Provide any information needed for the health and safety file.
	Inform the principal contractor of problems with the plan.
	Inform the principal contractor of reportable accidents, diseases and dangerous occurrences.

CDM co-ordinator duties

It is the CDM co-ordinator's duty to gather and prepare relevant information for the project.

CDM co-ordinator duties

All construction work (Part 2 of the regulations)	Additional duties for notifiable projects (Part 3 of the regulations)
	Advise and assist the client with their duties.
	Notify the HSE.
	Co-ordinate health and safety aspects of design work and co-operate with others involved with the project.
	Facilitate good communication between the client, designers and contractors.
	Liaise with the principal contractor about ongoing design.
	Identify, collect and pass on pre-construction information.
	Prepare and update the health and safety file.

A 02

Construction (Design and Management) Regulations

Duties of the designer

Designers are in a unique position, at an early stage of a project, to reduce the potential risks to health and safety that could arise during the construction phase. Building health and safety into the design element remains a key part of the regulations.

Designer duties

All construction work (Part 2 of the regulations)	Additional duties for notifiable projects (Part 3 of the regulations)
Eliminate hazards and reduce risks during design. Provide information about remaining risks.	Check that the client is aware of their duties and that a CDM co-ordinator has been appointed. Check that the HSE has been notified before starting work. Provide any information needed for the health and safety file.

As a result of legal interpretation, a designer is responsible for any drawings or design carried out by the staff within their employ and by staff they have under contract unless otherwise specified in writing under the terms of the contract.

Duties of all those involved

Finally, everyone involved with a construction project has a responsibility to look after themselves and others. Every person concerned in a project who is working under the control of another person shall report to that person anything which they are aware is likely to endanger the health or safety of themselves or others.

This includes:

- ☑ checking one's own competence*
- ☑ co-operating with others and co-ordinating work so as to ensure the health and safety of construction workers and others who may be affected by the work
- ☑ reporting obvious risks
- ☑ complying with requirements in Schedule 3 and Part 4 of the regulations for any work under their control
- ☑ taking account of and applying the general principles of prevention when carrying out duties.

Whilst there is no reference to 'developers' in the regulations, they will be regarded as a construction client. If there is any doubt as to how the regulations apply, advice should be sought from the Health and Safety Executive.

It is essential that everyone at every level in a construction project acknowledges their own limitations and at what point they must seek expert advice. Furthermore, extensive guidance on establishing and developing competence is contained in the HSE publication Managing health and safety in construction *(L144).*

The Construction (Design and Management) Regulations

Typical notifiable project under CDM

Project documents

There is a legal duty to compile the following project documents for notifiable projects only. However, the type of information gained by putting these documents together would be useful in ensuring the effective management of health and safety on some non-notifiable projects.

An example of what might be found in the three types of project document is described on the following page.

Construction (Design and Management) Regulations

Pre-construction information

It is the duty of the CDM co-ordinator to prepare the pre-construction information section in order to get the right information to the right people at the right time.

Description of project	Tick when completed/in place
1. Project description and programme details including:	
a) key dates (including planned start and finish of the construction phase)	
b) the minimum time to be allowed between appointment of the principal contractor and instruction to commence work on site.	
2. Details of client, designers, co-ordinator and other consultants.	
3. Whether or not the structure will be used as a place of work (in which case the design will need to take account of the relevant requirements of the Workplace (Health, Safety and Welfare) Regulations (as amended).	
4. Extent and location of existing records and plans.	
Client's considerations and management requirements	
1. Arrangements for:	
a) planning and managing the construction work, including any health and safety goals for the project	
b) communication and liaison between client and others	
c) security of the site	
d) welfare provision.	

Client's considerations and management requirements *continued*	Tick when completed/in place
2. Requirements relating to the health and safety of the client's employees or customers, or other people involved in the project, such as:	
a) the requirement for site hoardings	
b) site transport arrangements or restriction on vehicle movements	
c) client permit to work systems	
d) fire precautions	
e) emergency procedures and means of escape	
f) 'no-go' areas or other authorisation requirements	
g) any areas designated as a confined space by the client	
h) smoking and parking restrictions.	
Environmental restrictions and existing on-site risks	
1. Safety hazards, including:	
a) boundaries and access, including temporary access (for example, narrow streets, lack of parking, turning or storage space)	
b) any restrictions on deliveries, waste collection or storage	
c) adjacent land uses (such as schools, railway lines or busy roads)	
d) existing storage of hazardous materials	
e) location of existing services particularly those that are concealed (such as water, electricity and gas)	

Construction (Design and Management) Regulations

A 02

Environmental restrictions and existing on-site risks *continued*	Tick when completed/in place
f) ground conditions, underground structures or watercourses where this might affect the safe use of plant (such as cranes, or the safety of groundworks)	
g) information on existing structures – stability, structural form, fragile or hazardous materials, anchorage points for fall arrest systems (particularly where demolition is involved)	
h) previous structural modifications, including weakening or strengthening of the structure (particularly where demolition is involved)	
i) fire damage, ground shrinkage, movement or poor maintenance, which may have adversely affected the structure	
j) any difficulties relating to plant and equipment in the premises (such as overhead gantries whose height restricts access)	
k) health and safety information contained in earlier design, construction or as-built drawings (such as details of pre-stressed or post-tensioned structures).	
2. Health hazards, including:	
a) asbestos, including results of surveys (particularly where refurbishment and/or demolition is involved)	
b) existing storage of hazardous materials	
c) contaminated land, including results of surveys	
d) existing structures containing hazardous materials	
e) health risks arising from clients' activities.	

Significant design and construction hazards	Tick when completed/in place
1. Significant design assumptions and suggested work methods, sequences or other control measures.	
2. Arrangements for co-ordination of ongoing design work and handling design changes.	
3. Information on significant risks identified during design.	
4. Materials requiring particular precautions.	

The construction phase plan

It is the responsibility of the principal contractor to prepare the construction phase plan and provide contractors with relevant parts of the plan.

The level of detail should be proportionate to the risks involved in carrying out the work.

Description of project	Tick when completed/in place
1. Project description and programme details, including any key dates.	
2. Details of client, CDM co-ordinator, designers, principal contractor and other consultants.	
3. Extent and location of existing records and plans that are relevant to health and safety on site, including information on existing structures when appropriate.	
Management of the work	
1. Management structure and responsibilities.	
2. Health and safety goals for the project and arrangements for monitoring and review of health and safety performance.	

Construction (Design and Management) Regulations

Management of the work *continued*	Tick when completed/in place
3. Arrangements for:	
a) regular liaison between parties on site	
b) consultation with the workforce	
c) the exchange of design information between the client, designers, CDM co-ordinator and contractors on site	
d) handling design changes during the project	
e) the selection and control of contractors	
f) the exchange of health and safety information between contractors	
g) site security	
h) site induction	
i) on-site training	
j) welfare facilities and first aid	
k) the reporting and investigation of accidents and incidents, including near misses	
l) the production and approval of risk assessments and written systems of work.	
4. Site rules.	
5. Fire and emergency procedures.	

The Construction (Design and Management) Regulations

Arrangements for controlling significant site risks	Tick when completed/in place
1. Safety risks, including:	
a) delivery and removal of materials (including waste) and work equipment, taking account of any risks to the public (for example, during access to or egress from the site)	
b) dealing with services (water, electricity and gas), including overhead powerlines and temporary electrical installations	
c) accommodating adjacent land use	
d) stability of structures whilst carrying out construction work, including temporary structures and existing unstable structures	
e) preventing falls	
f) work with or near fragile materials	
g) control of lifting operations	
h) the maintenance of plant and equipment	
i) work on excavations and work where there are poor ground conditions	
j) work on wells, underground earthworks and tunnels	
k) work on or near water where there is a risk of drowning	
l) work involving diving	
m) work in a caisson or compressed air working	
n) work involving explosives	
o) traffic routes and segregation of vehicles and pedestrians	
p) storage of materials (particularly hazardous materials) and work equipment	
q) any other significant safety risks.	

A 02

Construction (Design and Management) Regulations

A 02

Arrangements for controlling significant site risks *continued*	Tick when completed/in place
2. Health risks:	
a) the removal of asbestos	
b) dealing with contaminated land	
c) manual handling	
d) use of hazardous substances, particularly where there is a need for health monitoring	
e) reducing noise and/or vibration	
f) work with ionising radiation	
g) exposure to UV radiation (from the sun)	
h) any other significant health risks.	
The health and safety file	
1. Layout and format.	
2. Arrangements for the collection and gathering of information.	
3. Storage of information.	
Significant design and construction hazards	
1. Significant design assumptions and suggested work methods, sequences or other control measures.	
2. Arrangements for co-ordination of ongoing design work and handling design changes.	
3. Information on significant risks identified during design.	
4. Materials requiring particular precautions.	

The Construction (Design and Management) Regulations

The health and safety file

It is the responsibility of the CDM co-ordinator to ensure that a health and safety file is prepared. The level of detail should allow the likely risks to be identified and addressed by those carrying out the work.

1. A brief description of the work carried out.	
2. Any residual hazards which remain and how they have been dealt with (for example, surveys or other information concerning asbestos, contaminated land, water bearing strata, buried services and so on).	
3. Key structural principles (for example, bracing, sources of substantial stored energy, including pre- or post-tensioned members) and safe working loads for floors and roofs, particularly where these may preclude placing scaffolding or heavy machinery there.	
4. Hazardous materials used (for example, lead paint, pesticides, special coatings that should not be burnt off and so on).	
5. Information regarding the removal or dismantling of installed plant and equipment (for example, any special arrangements for lifting, order or other special instructions for dismantling and so on).	
6. Health and safety information about equipment provided for cleaning or maintaining the structure.	
7. The nature, location and markings of significant services, including underground cables, gas supply equipment, fire-fighting services and so on.	
8. Information and as-built drawings of the structure, its plant and equipment (for example, the means of safe access to and from service voids, fire doors and compartmentalisation and so on).	

The file does not need to include things that will be of no help when planning future construction work, maintenance or demolition.

A 02

Construction (Design and Management) Regulations

Consultation with the workforce

The Approved Code of Practice, which supports the regulations, goes to great lengths to stress the vital importance of contractors, managers and supervisors engaging and consulting with the workforce as an effective way of identifying hazards and controlling the risks.

It is the workforce who have first-hand experience of actually carrying out the job and who may therefore be more knowledgeable about the risks involved.

Gaining willing and effective worker participation and feedback will be achieved when the workforce:

- ☑ has sufficient opportunity
- ☑ feels sufficiently confident in their managers and supervisors that their ideas and concerns will be listened to and, if required, acted upon
- ☑ has sufficient knowledge to recognise when something is not safe
- ☑ has been trained in the skills necessary to deliver effective feedback.

Ample opportunity must be given for all members of the workforce to consult with management on any concerns regarding health and safety that they might have.

Feedback boards provide opportunities for consultation

 For further information refer to Chapter A05 Leadership and worker engagement.

Competence

Throughout the regulations, the need for competence in the people working on the project (at all levels) is emphasised. Competence is seen as the key element in reducing accident figures and the incidence of occupational ill health that can arise out of construction activities. The regulations specify that duty holders must be sufficiently competent to:

- ☑ perform any requirement (under the regulations)
- ☑ avoid contravening relevant health and safety legislation.

The Construction (Design and Management) Regulations

The definition of *competence* is accepted as being a blend of skills, knowledge, attitude, training and experience (SKATE).

The Approved Code of Practice outlines in great detail how the competence of those involved in a project may be established or verified.

For further information refer to Chapter A03 Developing a health and safety management system.

Co-operation and co-ordination

Whether a project is notifiable or not, the regulation requires that all duty holders co-operate with each other and co-ordinate their work activities to enable the project to proceed without risks to health or safety.

An example of co-operation would be one contractor allowing the workers of another contractor to use a scaffold to carry out work that would otherwise have been done using a ladder.

Co-ordination of work activities should be an integral part of planning any project to ensure that, for example, the timing of any activity does not clash with any other so that nobody is put at risk.

A 02

Construction (Design and Management) Regulations

Health and safety on construction sites

Part 4 of the regulations outline the requirements for the management of health, safety and welfare on site. The following information is a summary of the duties of the principal contractor, contractors and the self-employed, with regard to health and safety on site.

Regulation number and title	Details of duties and requirements
25. Application of Regulations 26 to 44.	This regulation states that everybody carrying out construction work must comply.
26. Safe places of work.	Every place of work must be safe to access, egress and to work at.
27. Good order and site security.	Sites must be tidy, secure (to prevent unauthorised access), have appropriate signs and have no projecting sharp objects.
28. Stability of structures.	Buildings and structures must not be allowed to become unstable during work and any supports provided must be adequate for the job.
29. Demolition or dismantling.	These activities must be carried out so as to prevent any danger and there must be a written plan (for example, method statement) in place before work starts.
30. Explosives.	Explosives must be stored, transported and used safely.
31. Excavations.	Work in excavations must be properly planned. They must not be allowed to collapse or have materials falling or slipping into them. No-one should become trapped because of inadequate shoring. Excavations must be inspected and reports written as appropriate. Work in an excavation must not start where an inspection has revealed that it is not safe to do so.
32. Cofferdams and caissons.	Cofferdams and caissons must be suitably designed, constructed, maintained and inspected. Work in cofferdams or caissons must not start where an inspection has revealed that it is not safe to do so.
33. Reports of inspections.	The duty to carry out inspections refers to Regulations 31 and 32 and requires inspections to be made at the start of every shift and after any event likely to affect the safety of the excavation, cofferdam or caisson. A written record of an inspection must be made at intervals not exceeding seven days.

The Construction (Design and Management) Regulations

Regulation number and title	Details of duties and requirements
34. Energy distribution installations.	Energy distribution systems (power cables, pressure hoses/pipes and so on) above, on and below ground, must be protected against being damaged or causing injury or damage. Work must not take place where there is a risk of damage to, or accidental disturbance of, energy distribution systems, unless suitable control measures have been taken.
35. Prevention of drowning.	Where there is risk of a person drowning because of falling into water or other liquid, suitable measures must be put in place to prevent falls and, where a risk remains, to have a rescue plan in place. Where workers are transported over water to and from their place of work, suitable steps must be taken to ensure this is carried out safely.
36. Traffic routes.	Traffic routes must be organised to protect people from injury from vehicles, preferably by physical segregation, and be suitable for the vehicles and persons using them. Traffic routes must be properly maintained, suitably signed and regularly inspected. Where it is unsafe for pedestrians to use gates intended for use by vehicles, separate pedestrian gates must be provided and kept free of obstruction.
37. Vehicles.	Arrangements must be made to prevent unintended movement of any vehicle. They must be used safely, not overloaded and prevented from falling into any excavation. No-one must be allowed to ride on any vehicle unless it is designed to carry passenger(s) and it is safe to do so. No-one must remain on any vehicle that is being loaded unless a safe place is provided and maintained on the vehicle.
38. Prevention of risk from fire and so on.	Steps must be taken to avoid the risk of fire, flooding and asphyxiation.
39. Emergency procedures.	A plan must be in place for the safe evacuation of all people on site in situations where there is a foreseeable risk of an emergency. Everyone on site must be made aware of the emergency plan and it must be practised at suitable intervals.
40. Emergency routes and exits.	Adequate means of evacuation (emergency escape routes) to a safe place must be established. Routes must be kept clear of obstruction and be adequately signed.

A 02

Construction (Design and Management) Regulations

Regulation number and title	Details of duties and requirements
41. Fire detection and fire-fighting.	Suitable fire-fighting equipment, fire detection equipment and alarms must be provided, examined and tested at suitable intervals and suitably located. Manual fire-fighting equipment (such as hand-held extinguishers), must be easily accessible and indicated by appropriate signs. An appropriate number of on-site staff must be trained in the use of fire-fighting equipment.
42. Fresh air.	Places of work must have adequate fresh or purified air available. Where fresh or purified air is provided by plant, it must be equipped with suitable audible or visible warnings to indicate any inadequacy in the supply or machine failure.
43. Temperature and weather protection.	Steps must be taken to ensure that indoor temperatures are reasonable. Outdoor workplaces must be arranged to provide adequate protection against adverse weather, taking into account usage and any protective clothing or equipment provided.
44. Lighting.	Every place of work, its access and egress routes and any traffic route must be adequately lit, preferably by natural light. Where lighting is by artificial means, a back-up system must be in place where failure of the primary system would result in risks to health or safety. Artificial light must not adversely affect the perception of any health and safety sign.

03

Developing a health and safety management system

What your employer should do for you	48
What you should do as a supervisor	49
Competence in health and safety management	50
Basic principles	51
Setting your policy	51
Planning and setting standards	52
Organising and controlling staff	53
Consultation with employees	54
Measuring performance	55
Audit and review	56
Monitoring	57

Developing a health and safety management system

	What your employer should do for you
1.	Your company must provide a written safety policy (if more than five employees).
2.	The policy must be signed and dated.
3.	The policy needs to have a review date and it must be in date.
4.	Your company must identify the roles and responsibilities of everybody concerned.
5.	There must be a set of safety standards to be achieved.
6.	The set procedures and controls must be followed.
7.	There needs to be a procedure in place to measure performance and compliance.
8.	Systems and practices should be monitored, audited and reviewed.
9.	Your company must have a procedure in place for identifying training needs and providing the training.
10.	Measures must be in place to ensure effective two-way consultation and communication on matters of health and safety.
11.	Arrangements must be put in place to ensure effective co-operation between everyone who is or will be involved in a project.
12.	Your company should have a procedure for reporting and investigating near misses.
13.	Site inductions and other training provided for the needs of workers whose first language is not English, if required.
14.	The official Health and Safety Law poster must be displayed or each employee given a copy of the official leaflet or card.

What you should do as a supervisor

Checklist	Yes	No	N/A
1. Read, understand and comply with the company safety policy (if more than five employees).			
2. Check that the policy has been signed and dated.			
3. Check that the policy has a review date and is in date.			
4. Ensure that the company has identified the roles and responsibilities of everybody concerned.			
5. Know and understand your responsibilities as a supervisor.			
6. Assist with ensuring that safety standards are achieved as far as possible.			
7. Follow set procedures and controls, and report shortfalls.			
8. Monitor and report on practices, performance and compliance procedures.			
9. Contribute to audits and reviews carried out by the company or their health and safety advisor.			
10. Advise the company on training needs of yourself and your work team and assist with providing training if asked.			
11. Help to ensure an effective two-way consultation and communication on matters of health and safety.			
12. Co-operate with arrangements to ensure effective co-operation between everyone who is or will be involved in a project.			
13. Report near misses and co-operate with investigations.			
14. Ensure site inductions and other training provides for the needs of your work team, whose first language is not English as and when appropriate.			
15. Explain the official Health and Safety Law poster or a copy of the official leaflet or card to your work team.			

Competence in health and safety management

The effective management of health and safety within any company, large or small, will ultimately depend upon the competence of the person who takes on the responsibility for it.

 Competence is defined as capable and authoritative.

However, it is necessary to be more concise to enable employers to effectively target their efforts in achieving competence. The Concise Oxford Dictionary uses the words skills, knowledge, understanding and legal authority.

In health and safety circles, competence is often defined as a blend of:

Skills	your ability to manage and influence others
Knowledge	of health and safety issues
Attitude	your desire to achieve your (H&S) aims
Training	to gain the necessary knowledge/qualifications
Experience	can only be achieved over a period of time.

Health and safety law requires that every employer has access to competent health and safety advice, which in practical terms means that the employer must:

- ✓ be competent to manage relevant health and safety issues themselves, and/or
- ✓ have an employee who is, or
- ✓ bring in external expertise.

It is preferential that competence comes from within a company rather than be bought in through consultancy, which at best will only be on site sporadically. However, engaging a consultant in the short term might be essential whilst in-house competence is being gained.

With sufficient effort any employer, or the employee to whom the responsibility is delegated, can become sufficiently competent to manage health and safety issues within their company. It should be less demanding to achieve competence in managing the health and safety issues relevant to smaller companies, which repeatedly carry out a broadly similar type of work, although achievement is no less important than for a larger company. Competence is often demonstrated by the award of qualifications or the certification of training received.

No single event is going to make someone fully competent in health and safety but taking the first step is important. Reading articles, books and so on, attending courses, joining local health and safety groups or trade associations can all contribute to gaining competence. Studying the relevant parts of this book is one step along the way.

Competence must be established on two levels, corporate (company) and individual.

Corporate competence is the ability of any company to manage health and safety effectively through its policies and procedures.

Individual competence is the ability of every person to work safely without endangering themselves or anyone else and to be able to recognise if an unsafe situation is arising.

To establish your competence you will need to comply with the following points.

Developing a health and safety management system

- [✓] **As a contractor** on a non-notifiable project (for example, for a domestic client or for work that will last for less than 30 days), you will have to satisfy yourself of the competence of yourself, your employees and any other people (such as sub-contractors), that you bring on to site so that you can prove your and their competence, should you need to.

- [✓] **As a contractor on a notifiable project** you will have to satisfy the principal contractor of the competence of yourself, your employees and any other people (such as sub-contractors), that you bring on to site. Even on non-notifiable jobs you should still assure yourself of these facts so that, should you need to, you can prove both your and their competence.

- [✓] **As principal contractor** you will have to satisfy the requirements and enquiries of the client and CDM co-ordinator with regard to you being able to satisfactorily discharge the required duties. Amongst other things this will require you to take into account the competence of contractors and self-employed persons who are tendering for work on the job, and that of anyone else that they will be bringing on to site.

Basic principles

The HSE recommends a step-by-step approach consisting of:

- [✓] setting your policy
- [✓] planning and setting standards
- [✓] organising and controlling staff
- [✓] measuring performance
- [✓] audit and review.

To summarise, the basic principles of a health and safety policy means that a company must have:

- [✓] a **health and safety policy statement** that outlines the company's intentions
- [✓] an **organisational structure** that shows the responsibility tree (some larger companies may have one on each site)
- [✓] the **arrangements** that explain the duties of each position on the responsibility tree.

In many companies there is also an **environmental policy** *(refer to Chapter E23 Environmental management).*

 Health, safety and the environment must be proactively managed, in the same way that a good supervisor and manager will manage people, quality and productivity.

Setting your policy

From the summary above it can be seen that the essential starting point for all health and safety must be the preparation of a policy statement. This will inform employees of the company's intentions towards the health and safety of the workforce and the standards to which they aspire.

Health and safety at work is a legal requirement and must always be an integral part of any business activity.

Safety policy (policy, organisation and arrangements)

Every employer with five or more employees is required to have a written statement of their general policy on health and safety at work, together with details of the organisation ('the who') and arrangements

Developing a health and safety management system

necessary to put the commitment into practice ('the how'). Employers must ensure it is clearly displayed or give a copy of it to each employee and ensure that it is fully understood.

Even employers with fewer than five employees must formulate how they are going to proactively manage health and safety as an integral part of their work activities.

There is no reason why a company with fewer than five staff should not write its own health and safety policy. In seeking business from larger companies, it may be very useful – or essential.

A policy should describe in full:

- ☑ how the company is going to set up and maintain a safe working environment
- ☑ what health and safety responsibilities already exist
- ☑ who is responsible and for what
- ☑ the arrangements for developing safe systems of work
- ☑ arrangements for the review and update of the policy as necessary.

The policy should be signed and dated by the person with ultimate responsibility for health and safety. It should be periodically reviewed and updated as necessary, then re-signed and dated.

Many large companies and organisations (such as Local Authorities that offer contract work by way of tender), will only include companies in their approved list of tenders if they have been supplied with a copy of the prospective contractor's safety policy statement and if they have approved it.

Planning and setting standards

Nobody can develop or implement plans, or set standards, until they are sufficiently competent in health and safety to know what the law requires to be done and how to interpret those requirements into safe and healthy working practices.

The aim of any health and safety management system must be to prevent injuries, occupational disease and near-misses from occurring. Several indicators (such as the building and construction industry's ongoing accident rate and the growing list of HSE prosecutions), show that this simply will not happen unless plans and standards are put in place and enforced from within. In doing this, employers must:

1. commit themselves to proactively managing the health and safety aspects of their work activities
2. actively involve employees in the health and safety decision making process and maintain their interest
3. establish areas of health and safety weakness
4. select appropriate actions to overcome these weaknesses
5. ensure that sufficient resources (time, competent people and money) are in place to implement the actions
6. implement the actions and make sure they are complied with
7. monitor the results to establish if improvements can be made
8. go back to point 2 and follow through for any aspects where further improvements are required
9. investigate when things go wrong to prevent them happening again.

Developing a health and safety management system

Ultimately, the effectiveness of the plans will depend upon the assessment and control of the risks to health and safety, which arise out of the employer's work activities. As expanded upon later in this book, the Management of Health and Safety at Work Regulations place a legal requirement on employers and the self-employed to carry out suitable and sufficient assessments of the risks to the health and safety of employees and other people who may be at risk. *(Further details of assessment programmes may be found later in this chapter.)*

Risk assessments must be reviewed periodically to ensure they are still valid and be amended, if necessary.

Organising and controlling staff

Who does what?
Does everybody know what their specific responsibilities and roles are with regard to maintaining safe places of work?

Who is going to:
- [✓] develop and implement safe systems of work?
- [✓] issue instructions on how work is to be carried out safely?
- [✓] supervise or manage day-to-day health and safety issues?
- [✓] ensure everyone is competent to do what is required of them?
- [✓] develop risk assessments and (if necessary) method statements?
- [✓] carry out periodic health and safety inspections?
- [✓] deal with maintenance matters?
- [✓] measure health and safety performance and review/update procedures as necessary?

The Management of Health and Safety at Work Regulations and the Construction (Design and Management) Regulations specify that any employer, contractor (or principal contractor) who appoints anyone else to carry out work is responsible for checking the appointee's resources and competence, and ensuring that they have developed and adhere to safe systems of work.

Training
The Health and Safety at Work Act 1974 requires that employers provide all necessary information, instruction, supervision and training to enable their workforce to carry out their tasks safely and without risks to their health.

 Questions that should be asked:
- [✓] Is there provision for training now, and for the future?
- [✓] Who is going to identify the gaps between the skills needed and those held to establish the training needs of individuals and be responsible for ensuring that it is carried out?

Training may take many forms in addition to conventional training room sessions. Some of these are:
- [✓] site inductions
- [✓] toolbox talks
- [✓] safety leaflets, books and information on posters
- [✓] reading the company's health and safety policy
- [✓] on-the-job training and instruction on particular plant, tools or equipment
- [✓] having risk assessments or method statements explained.

Developing a health and safety management system

Procedures should be in place to:

- [✓] track what training has been received by whom
- [✓] provide assessment and feedback to confirm that the training was appropriate and has been understood and, if not, arrangements made for retraining.

Everyone needs to be involved to ensure a safe site

Communications

What are the arrangements for:

- [✓] enabling anyone to report suspected or actual failings in health and safety management to someone in authority?
- [✓] reporting and recording accidents?
- [✓] reporting dangerous occurrences or near misses?
- [✓] receiving and distributing information?
- [✓] delegating health and safety duties?
- [✓] effectively communicating with workers whose first language is not English?*

* Refer to Chapter A05 Leadership and worker engagement for further guidance on communicating with workers who have limited or no understanding of English.

The system of communication on larger sites, with perhaps many companies on site and several tiers of sub-contractor, must allow for the health and safety concerns of individuals to be considered by someone with the authority to rectify the situation. It is quite possible for the person on the tools and their supervisor to be more aware of an unsafe situation developing at the place of work. Employees must be given the opportunity to feed any health, safety or welfare concerns that they have directly to a site-based manager or supervisor who has the authority to take the appropriate action.

If you (or your contractor) employ union members then be prepared to accept their role in inspections and safety committees as described below.

Consultation with employees

An essential part of health and safety management is the two-way communication process between employers (or their representatives) and employees. There are legal duties on both parties with the intent that communication takes place in both directions.

 Employers and their managers must talk to, and listen to, the people who actually do the work.

Developing a health and safety management system

Where recognised trade unions are present within companies, they may appoint safety representatives on behalf of their members. The purpose is to create a situation where co-operation between the employer and the workforce is maximised.

The Safety Representatives and Safety Committees Regulations give certain functions to safety representatives. The prescribed functions include:

- ☑ carrying out workplace safety inspections
- ☑ involvement in accident investigations
- ☑ taking part in safety committees.

Large sites may provide customised site safety guides

Where there is no union representation, the requirements of the Health and Safety (Consultation with Employees) Regulations apply. These place a legal duty on employers to engage in two-way communication on matters of health and safety with their employees, either through direct contact or through the employees' representatives. As part of your consultation process, you must display the official Health and Safety Law poster in places where it is readable by all employees, or give each employee a copy of the official leaflet or card containing the same information.

 For further information refer to Chapter A05 Leadership and worker engagement.

Measuring performance

There are several ways in which a company may measure the effectiveness of its health and safety management system. They include:

- ☑ evaluating the effectiveness of earlier risk assessments
- ☑ gathering and evaluating information on the number of accidents over a reference period, their causes and effects
- ☑ the investigation of worker feedback on near misses and dangerous occurrences
- ☑ gathering information on reportable diseases
- ☑ the results of health and safety audits
- ☑ the results of inspections, whether regular, snap (unannounced) or specific
- ☑ the level of health and safety awareness among employees in the form of feedback from training courses and safety committee meetings
- ☑ benchmarking against previous audits.

Developing a health and safety management system

Gathering information regularly makes performance measurement easier

The HSE has a free leadership and worker involvement toolkit aimed at reducing harm by learning from the best in the construction industry. The toolkit has been developed by the construction industry to help contractors, managers, supervisors and workers learn how to make health and safety improvements in their businesses.

The toolkit is available at www.hse.gov.uk/construction/lwit

Audit and review

The process of carrying out health and safety audits, evaluating the findings and implementing remedial action, where necessary, results in a continuous cycle of improvement.

Audits tend to be formal, pre-planned, events with the findings documented for later evaluation and review.

By comparison, health and safety inspections are normally a less formal style of audit; the results are not necessarily recorded and they may be more appropriate on some smaller sites. Remember, however, no record – no proof!

Both health and safety audits and inspections are carried out on a pre-notified or no-notice basis as suits the situation at the time. If too many are pre-notified or regularly scheduled, it may tend to create a false impression, as those on site may only improve their behaviour or practices in the short term.

The important thing is that, where shortcomings are found, remedial action is promptly taken and lessons learned to prevent recurrence.

Safety audits

A safety audit is a demonstration of the management's commitment to monitor and improve, where necessary, the effectiveness of the company's safety management system from the top down. Thorough audits may discover health and safety failings that arise through factors that occur off site (such as design issues), which are therefore outside the direct control of site-based staff. They may also identify good practices that should be shared and incorporated into a process of improvement.

Developing a health and safety management system

Audits should be carried out by a member of management, often the company health and safety adviser or a director, who may or may not be accompanied by a representative(s) of sub-contractors or the employees. The aim is to identify problem areas, implement improvements and increase the overall standard of safety awareness.

It requires attention to the:

- ✓ work environment
- ✓ person carrying out the task
- ✓ task itself
- ✓ safety culture
- ✓ way in which each affects the others.

CITB has developed the *Construction site health, safety and environment auditing system* (SA 03 CD).

The system is in two parts.

- ✓ Health, safety and environmental management systems audit dealing with auditing the procedures and arrangements upon which the health and safety management system is based.
- ✓ Site operations audit covering the way in which company policies and procedures are put into practice on site.

Monitoring

Employers are required to monitor performance and there are two types of monitoring used in auditing and inspecting.

- ✓ **Proactive monitoring** measures the current level of compliance with legislation and company procedures.
- ✓ **Reactive monitoring** investigates anything that has actually happened: accidents, incidents, near-misses and ill health.

Health and safety inspections, including recording and reporting (proactive monitoring)

Health and safety inspections tend to be the site manager's snapshot of the health and safety standards on site at any one time. Alternatively, a company's health and safety adviser or director may carry out these inspections.

A safety inspection may examine the big picture or concentrate on only one feature; irrespective of who carries out the inspection, they must have the knowledge to enable them to detect unsafe situations and the authority to ensure they are rectified.

Investigating accidents, incidents and ill health, including recording and reporting (reactive monitoring)

Accidents and illness resulting in serious injuries and fatalities, including health hazards, are too often a feature of work in the construction industry. The proactive management of health and safety should serve to keep such incidents to a minimum. However, the situation needs to be managed should an accident or incident occur.

It should not be forgotten that a near miss today could be a potential accident tomorrow. Supervisors have an important role to play in both proactive and reactive monitoring activities.

For further information on accident prevention, recording and reporting refer to Chapter A07 Accident reporting and emergency procedures.

Developing a health and safety management system

A 03

04

Risk assessments and method statements

What your employer should do for you	60
What you should do as a supervisor	61
Introduction	62
The principles of risk assessment	63
The stages of risk assessment	64
Establishing control measures	66
Young persons	67
New and expectant mothers	68
Permit to work	68
Method statements	68
Point of work risk assessment for supervisors	69
Risk assessment tools	69
Calculating the risk rating	70

Risk assessments and method statements

A 04

What your employer should do for you
1. Identify and assess workplaces, work tasks and procedures.
2. Provide suitable assessors to carry out assessments.
3. Ensure assessors have the necessary knowledge, experience or qualifications (competence) to carry out suitable and sufficient assessments.
4. Arrange for training if required, and put arrangements in place and nominate a person to be responsible.
5. Select and implement a suitable risk assessment and method statement process.
6. Put arrangements in place to communicate the significant findings of the risk assessments.
7. Appoint a person with the authority to decide upon and implement any control measures that are considered necessary.
8. Establish assessment review dates.
9. Assess the needs of young persons, if employed.
10. Have a policy in place for the assessment of new and expectant mothers.
11. Put procedures in place to ensure the safety of workers to ensure communication is effective (for example for those whose first language is not English or have reading or writing difficulties).

Risk assessments and method statements

What you should do as a supervisor

Checklist	Yes	No	N/A
1. Ensure workplaces, work tasks and procedures are safe for your work team and others who may be affected.			
2. Assist assessors in preparing risk assessments and method statements.			
3. Provide practical knowledge and experience to assessors to ensure assessments are suitable and sufficient.			
4. Identify if training is required, and report shortfalls to a person nominated to be responsible in the company.			
5. Check that the risk assessment and method statement match the tasks to be carried out.			
6. Communicate the findings of the risk assessments to your work team and others who may be affected.			
7. Check that any control measures that are considered necessary are implemented.			
8. Ensure that the risk assessments and method statements are current and match the workplace situation.			
9. Take extra care of young persons and recognise their shortfalls due to immaturity and lack of experience.			
10. Implement the company policy for new and expectant mothers, where applicable.			
11. Ensure your workers whose first language is not English, or who have reading or writing difficulties, have a clear understanding of their work and emergency measures are in place on site.			

A 04

Risk assessments and method statements

Introduction

The principle of risk assessment is a fundamental cornerstone of the management of health and safety in the workplace. The Management of Health and Safety at Work Regulations (The Management Regulations) place a legal duty on **employers** to assess the risks, that arise out of their work activities, to the health and safety of:

- ☑ their employees
- ☑ any other people who are not their employees.

The regulations place a similar duty on the **self-employed** to safeguard the health and safety of themselves and anyone else who may be affected by their work activities.

The regulations also require that arrangements are made for the effective planning, organisation, control, monitoring and review of the preventive and protective measures found necessary to control the risks to health or safety identified by a risk assessment. If there are five or more employees, the significant findings of each assessment must be recorded.

The requirement for risk assessments also extends to a company's office premises, storage yards and so on, if they exist, plus the use of vehicles on company business.

Most of the chapters in this book will highlight common types of work activity that are inherently unsafe if the risks to health and safety are not adequately controlled. This chapter explains the principles and process of risk assessment for controlling such risks.

In addition to the above requirement for general risk assessments, there is also a requirement in other sets of regulations for employers to carry out risk assessments in relation to specific threats to health or safety in the workplace, such as:

- ☑ the use of hazardous substances (COSHH)
- ☑ noise in the workplace
- ☑ hand-arm vibration and vibration white finger
- ☑ manual handling activities
- ☑ the presence of asbestos
- ☑ exposure to lead.

However, this does **not** put an obligation on employers to carry out two risk assessments for the same hazard. If a general risk assessment for a work activity adequately covers the specified hazards, that risk assessment alone will be sufficient.

There is no legal definition of a *risk assessment* but, in practice, it can be described as:

> **a careful and structured examination of a work activity (or a group of associated work activities) to identify any features of the work that could harm the health or safety of anyone and how, by the implementation of effective control measures, the risk of harm occurring may be eliminated or reduced to an acceptable level.**

If a risk assessment is to be effective, it is essential that the person who carries it out is familiar with all aspects of the task being assessed. It might be found beneficial in some cases to involve the workers who are familiar with carrying out the type of work, in order to gain a better insight into any problem areas. Risk assessments should focus on what is known to actually happen on site rather than what should happen.

Risk assessments and method statements

Whilst risk assessments are primarily about reducing injuries, deaths and occupational ill health, the damage to plant and equipment, and the avoidance of environmental harm, should also be considered during the risk assessment process.

In addition to the legal requirement, written risk assessments could provide considerable business benefits. Lower accident rates mean less time off work for employees (and therefore better business continuity), lower insurance premiums and possibly getting onto preferred contractors' lists and repeat work from satisfied clients.

Some terms used in risk assessment

Anyone carrying out risk assessments must be familiar with the meaning of the following terms that are used in the process.

Hazard: anything that has the potential to cause harm (ill health, injury or damage).

Risk: the likelihood of an event occurring from a hazard.

Likelihood: what is the chance that an accident will occur (certain, likely, possible, unlikely or rare)?

Severity: what will be the severity (consequences) of any incident that arises?

Danger: a person is in danger when they are exposed to a risk.

Accident: an event that results in injury or ill health.

Near miss: (including dangerous occurrence) an event that, while not causing harm, has the potential to cause injury or ill health.

Competence: having practical and theoretical knowledge, training and actual experience of the work activities involved.

 Severity: at one extreme, someone who is buried in a collapsed excavation could be killed, whereas the consequences of someone not using a hand tool correctly might result in a cut or grazed skin.

The principles of risk assessment

All recent health and safety legislation is structured around a goal setting or risk-based approach to the management of health and safety in the workplace. Prescriptive control measures (such as the minimum height of scaffold guard-rails), only occur in a few instances.

The risk assessment cycle is:

- ☑ **identify the hazards** that arise out of the work activity being assessed
- ☑ **eliminate the hazards** (where possible) thereby removing the risk of injury
- ☑ **assess the risks** to the health and safety of any person(s) arising out of the remaining hazards
- ☑ **identify the individuals** or groups of people who are at risk
- ☑ **record** the significant findings of the risk assessment
- ☑ **monitor and review the control measures** established for the work activity and amend them if they are no longer valid or become ineffective.

Risk assessments and method statements

 Particular provision is made in the regulations for the protection of young persons (under 18 years of age) and some female employees in the workplace.

The controls on what young people are allowed to do in the workplace are quite strict and specific. A full outline of the factors that must be considered when risk assessments have to take account of young people, or some female employees, is included in this chapter.

The stages of risk assessment

1. Look for the hazards.
Some typical examples found on construction sites are:

- ☑ an untidy site with lots of slipping and tripping hazards
- ☑ the use of equipment with rotating blades (such as disc cutters)
- ☑ the use of power tools, creating dust
- ☑ the risk of fire from spark-emitting tools (such as angle grinders)
- ☑ work at height with the potential for falls (for example, roofing activities)
- ☑ manual handling activities or working in cramped conditions
- ☑ the operation of construction plant near to people on foot
- ☑ the presence of contaminated ground
- ☑ the use of chemicals, solvents, paints and so on.

2. Decide who might be harmed and how.
Having established the hazards associated with the job, you will have to identify the individuals or groups of people who are at risk, for example:

- ☑ yourself
- ☑ other employees (particularly anyone young and inexperienced)
- ☑ employees who need special consideration (for example, someone who is deaf or who is not fluent in English)
- ☑ female employees who are pregnant, or nursing mothers
- ☑ employees of other contractors
- ☑ visitors (such as delivery drivers, maintenance staff and clients)
- ☑ members of the public or trespassers (particularly children)
- ☑ anyone else who might be affected by your work (for example, neighbouring businesses).

3. Evaluate the risks and decide upon precautions.
Having identified the hazards, now consider what risk-control measures are necessary to keep from harm the people who have been identified as being at risk. Even when precautions have been put in place there will usually be a remaining (residual) risk. If you choose to use the risk assessment tool included in this chapter, you will have to exercise your judgment as to whether, in your opinion, the residual risk is **high,** medium **or** low**.**

This part of the assessment is a loop. If, after identifying risk control measures, you think that the level of residual risk is still too high, it will be necessary to introduce further control measures, again evaluate the level of residual risk and continue the process until it is considered to be acceptable. It is a case of you asking yourself:

Risk assessments and method statements

 What have I done to control the risks and what more do I need to do?

Using a remote controlled trench compactor eliminates the risk of a person entering the excavation

4. Record your findings and implement them.

Health and safety law requires that all employers carry out risk assessments for their work although, legally, only employers with five or more employees need record the significant findings of their risk assessments. However, in practice, many of the larger contractors will not accept any sub-contractor on site, even those with fewer than five employees, if they do not have written risk assessments. Risk assessments may be recorded electronically, providing that they may be easily retrieved for scrutiny if required.

In the interests of clarity, risk assessments may identify where relevant supporting information can be found (such as the company health and safety policy), rather than including masses of information in the risk assessment itself. However, if this approach is taken, the supporting documents may also have to be supplied to any person who wishes to see the risk assessment.

The law does not specify how a risk assessment should be laid out, although two common conventions have evolved over time.

- ✓ **A quantitive risk assessment,** which outlines the hazards present and the risk control measures necessary, and in which residual risk is given a numerical score or categorised as 'high', 'medium' or 'low' to enable corrective actions to be prioritised.

- ✓ **A qualitative risk assessment,** in which the process is the same. No attempt is made to quantify the level of residual risk.

Examples of both types of risk assessment are included in this chapter.

 Irrespective of how a risk assessment is laid out, it is considered good practice for it to identify the:

- ✓ **person who will manage the residual risks**
- ✓ **date(s) by which any essential actions must be taken, in the interests of health or safety.**

5. Review the assessment and revise if necessary.

Risk assessments should be reviewed from time to time to ensure the control measures are still appropriate and effective. If there is a change to any aspect of the way the job has to be carried out, that might affect health and safety, for example:

- ✓ having to use a different item of equipment part way through the job

Risk assessments and method statements

- the arrival of a new operative who is inexperienced
- unexpected deteriorating weather
- the late arrival of materials.

The assessment must be reviewed and the risks re-evaluated.

Suitable and sufficient (employer duties) – summary

The regulations require that all risk assessments are suitable and sufficient. For a risk assessment to comply with this requirement, it must:

- establish the risks arising from the work activity
- be appropriate, given the nature of the work, and such that it remains valid for a reasonable period of time
- be proportionate to the level of risk and the nature of the work
- identify and prioritise the control measures required to protect the health and safety of the employees and others who may be affected.

Suitable and sufficient (supervisor duties) – summary

- Receive, if necessary question, and accept the risk assessments.
- Check the suitability of the assessments when compared with the workplace (carry out a point of work risk assessment).
- Question anything that does not seem to be right. **If in doubt ask!**

Establishing control measures

When establishing appropriate control measures, consideration should be given to the following techniques.

- **Combat risks at source** (for example, use a safe product rather than rely on personal protective equipment (PPE)).
- **Take advantage of** technical progress and adopt new, safer methods of working (for example, use a modern trestle system, complete with guard-rails and toe-boards, rather than scaffold boards supported on improvised hop ups).
- **Replace the dangerous with the non-dangerous or less dangerous** (for example, prohibit the use of all 230 V power tools, allowing only 110 V or battery-powered tools).
- **Adopt measures that protect the greatest number of individuals** (for example, safety nets protect everyone working above them, whereas a safety harness only protects the wearer).
- **Give appropriate information**, instructions and training to employees and others – assess the need for training.
- **Provide PPE: always the very last resort.**

What do you do after your risk assessment as an employer?

- Put into effect the measures that you have decided will adequately control the risks.
- Communicate the findings of your risk assessments, particularly details of the hazards identified and what control measures are in place, to anyone who needs to know (for example, the main (or principal) contractor, your employees, sub-contractors and so on).

It is most important that the findings of risk assessments are communicated to anyone whose health and/or safety is likely to be affected by the job. Do not just store them away.

Risk assessments and method statements

What do you do after you receive a risk assessment as a supervisor?

- ☑ Speak to the site manager on a regular basis.
- ☑ Check that the risk control measures to be put in place are right for the work situation.
- ☑ Seek clarification of uncertainties and make your manager aware of any findings of a point of work risk assessment so that control measures are corrected.
- ☑ Before briefing your work team, ensure the risk assessments and method statements are correct.
- ☑ Explain the risk assessments and method statements to your work team, ensure understanding (by asking questions) and record the names of those briefed (and when).

> **!** Do not rely on generic risk assessments. As a supervisor it is your responsibility to make sure that your workplace is safe.

Young persons

A young person is classified in law as anyone who is not yet 18. A risk assessment for a work activity in which a young person will be involved must take account of:

- ☑ their inexperience, lack of awareness and immaturity
- ☑ the layout of the workplace (for example, will young people be expected to work at height or in excavations)
- ☑ the nature, degree and duration of exposure to physical, biological or chemical agents (noise and vibration are physical agents)
- ☑ the type of equipment that the young person would be expected to operate and how work is organised/supervised
- ☑ the amount of health and safety training that the young person has received
- ☑ possible exposure to extremes of hot or cold.

In practical terms, the main anticipated implications of there being a young person on site are restrictions in what they are allowed to do and, initially at least, a higher level of supervision.

These bullet points are not intended to discourage the employment of young persons; they only highlight the factors that must be taken into consideration when planning the work that they are going to do.

Any risk assessments carried out for young people under the age of 16 at work (for example work-placement trainees) must be shared with their parents or guardians.

Special arrangements must be in place for young persons on site

Risk assessments and method statements

New and expectant mothers

Special consideration must be given within risk assessments to any employee who is pregnant, has given birth within the past six months or who is breastfeeding, where the nature of the work would put her at risk because of her condition.

The assessment may identify the need to alter the work environment, work pattern, work activity, working hours, and provide additional temporary facilities or support.

Any employee who is pregnant, has given birth within the past six months or who is breastfeeding must notify her employer in writing of her condition within a reasonable length of time.

Permit to work

Whilst there is no requirement in law to use a permit to work system, they are often used to regulate how potentially high risk activities are to be carried out in a healthy and safe manner. As such, they very much support the risk assessment from which they are derived.

Permits to work are often, but not exclusively, required where the work:

- involves entry into a confined space
- depends upon the isolation of high-voltage electrical equipment
- involves the disturbance of any system carrying a fluid or gas under pressure
- involves hot works.

A permit to work is a formal, dated and time-limited certificate signed by a properly authorised and competent person. Receipt of a permit is acknowledged by the signature of the person in charge of the work, who will retain a copy; another copy will be retained by the person issuing the permit to work.

Other signatures on the permit will certify that any control measures necessary for the job to start have been implemented (for example, locking-out an electrical supply or checking the atmosphere in a confined space).

A permit will indicate the time and date at which it expires. If the work is not completed at that time, depending upon the circumstances:

- it is usually necessary to make everything safe, for everyone to leave the work area before the permit expires and for the permit to be cancelled
- it may be safe to extend the expiry time and for the work area to be reoccupied.

When the work is completed, or the expiry time has passed, the person in charge of the work must return their copy of the permit to the person who issued it and the permit is cancelled.

Method statements

A method statement is a document that describes in a logical sequence exactly how a work activity is to be carried out in a manner that is safe and without risk to health. They enable the principal (or main) contractor to examine the proposed working methods of all contractors and sub-contractors and to establish where jobs that are to take place at the same time may conflict with regard to the interests of health and safety. Well written method statements provide an ideal means of communicating vital health and safety information to those who will be doing the work, usually by way of the method statement being explained to them.

To a large extent, the way in which any job will be undertaken, and therefore how it is detailed in the method statement, will reflect the findings of the risk assessment(s) for the same job. The control measures selected for controlling risk will influence the method of carrying out the job. In most cases, the person writing the method statement will have to extensively refer back to the risk assessment(s) for the same job.

For routine and repetitive activities (work that is carried out many times where the hazards and risk are broadly the same), a previous method statement may be applied again provided it has been reviewed at point of work (by the supervisor) to ensure that it is still relevant. This is referred to as a **generic method statement**. Where the work is new, more complicated or unusual, then a specific method statement will have to be produced.

Point of work risk assessment for supervisors

An employer's risk assessment has generally not been prepared **on the day at the workplace** so supervisors must ensure that they are correct before briefing their work team. A simple point of work risk assessment could be used.

Remember the STAR principle	☑ Stop
	☑ Think
	☑ Act
	☑ Review

Before you start work:

- ☑ **Stop** and **think** about where you are and what you have to do.
- ☑ If it's safe to proceed you can start, or else make it safe (**act**).
- ☑ When you have finished **review** what you have done.
- ☑ If anything was learnt from the job, report back to your employer (lessons good or bad for improvement in the future).

 Think before you act – has the situation changed? If in doubt, stop and ask.

Risk assessment tools

Qualitative risk assessment

The risk of something going wrong is considered in terms of **likelihood** (probability) and **severity** (consequences).

The **likelihood** of a hazard actually causing harm or an accident is rated as being **high,** medium or low in accordance with the following.

High   It will happen regularly, or it could be a usual or a common occurrence.

Medium   It is less regular, but is still recognised as being likely to happen.

Low   It has not happened for a long time; it is known to be infrequent and is not likely to happen.

Risk assessments and method statements

The severity of the event, should it happen, can then be categorised as follows.

High   The result could be a fatal accident or multiple injuries/major property damage/substantial pollution or environmental impact.

Medium   It would probably cause serious injuries, or persons would be off work for over seven days due to their injuries/substantial property damage/there may be some pollution.

Low   There would be minor injuries to persons or some slight damage to property.

Calculating the risk rating

1. Likelihood and severity are mapped on a matrix, as shown below.

Likelihood	High			
	Medium			
	Low			
		Low	Medium	High
		Severity		

2. As an example, for an activity where likelihood is assessed as **high** and severity is assessed as **medium**, the overall risk rating is plotted, as shown below.

Likelihood	High		X	
	Medium			
	Low			
		Low	Medium	High
		Severity		

3. The combined likelihood/severity risk rating is then graded by taking the highest of the two individual ratings, as shown below.

Likelihood	High	**High**	**High**	**High**
	Medium	Medium	Medium	**High**
	Low	Low	Medium	**High**
		Low	Medium	High
		Severity		

- ✓ A combined risk rating of **high** should be totally unacceptable and the work should not be undertaken until the risk has been reduced.
- ✓ When there is a combined risk rating of **medium**, action must be taken and work stopped, if necessary, to reduce the risk level.
- ✓ If the combined risk rating is **low**, it is acceptable to start the work as long as everything reasonably practicable has been done in order to reduce the risk, and the assessment is reviewed at regular intervals.

Using this information, decisions can now be made on whether it is sufficiently safe to continue with an activity, or whether further control measures are necessary.

Risk assessments and method statements

Consider this example

Roof trusses have to be installed on a two-storey house. One of the obvious hazards is a fall from height. If a ladder is to be used to access the top of the walls with no further protective measures taken, the likelihood of a fall back to the ground during the installation of the trusses could be assessed as **high**. Similarly, the severity of a fall from height, particularly someone falling from eaves height onto the floor below, could also be assessed as **high**.

If instead, a scaffold platform were erected around the house at eaves height and airbags installed below across the full span of the house, both the likelihood of a fall occurring, and the severity of any fall that did occur (onto the scaffold platform or onto the airbags), could be assessed as low.

By revising the work method, a **high**/**high** situation, which gives a combined risk rating of **high**, has been reduced to low/low, which gives a combined risk rating of low.

An alternative work method, by which the whole roof-truss assembly is manufactured at ground level and craned into place, would also serve to reduce the risks of working at height.

(A suggested format for laying out risk assessments can be found on the next three pages.)

Example of a quantitive risk assessment

Company	
Project title	
Location	

Assessment date	Review date	Name/contact details of assessor	Unique reference

Work activity	Working in and around excavations.
Persons at risk	Operatives carrying out the job, other contractors and site visitors.

Risk assessments and method statements

Hazard No. 1	Fall of people or plant into the excavation.		
Risk control measures	Safe ladder access into and out of excavation. Install fencing around excavation. Install lighting. Install warning signs. People/vehicles not involved in activity excluded from the area. Stop blocks installed to keep site vehicles at a safe distance. Construction plant tipping into the excavation controlled by a trained signaller.	*Owners of the risks*	*Date by which any actions must be taken*

Risk rating	High	Medium	Low	Combined risk rating
Likelihood			X	Medium
Severity		X		

Risk assessments and method statements

Hazard No. 2	Collapse of sides/falling materials.				
Risk control measures	Side supports installed by competent persons. Statutory inspections carried out. Side supports not used as means of climbing into or out of excavation. Number of operatives allowed in excavation kept to a minimum. No adverse weather forecast that is likely to affect ground conditions during job. Spoil piled a safe distance from top of excavation.			*Owners of the risks*	*Date by which any actions must be taken*
Risk rating	**High**	**Medium**	**Low**	**Combined risk rating**	
Likelihood			X	Medium	
Severity		X			

Note: the Construction (Design and Management) Regulations require that all practicable steps shall be taken, where necessary, to prevent danger to any person. This includes, where necessary, the provision of supports or battering.

A 04

Risk assessments and method statements

Hazard No. 3	Accidental contact with underground services.					
Risk control measures	Pre-excavation ground survey carried out. Utility companies contacted and attended the site. Site plans examined. The only known underground service is a drain carrying rainwater.				*Owners of the risks*	*Date by which any actions must be taken*
Risk rating	**High**		**Medium**	**Low**	**Combined risk rating**	
Likelihood				X	Low	
Severity				X		

Risk assessments and method statements

Example of a qualitative risk assessment

Activity assessed	Delivery of bulk construction materials.		Page 1 of 1
Company name		Assessment date	
Company address		Assessment by *(print name)*	
Site address		Assessment review date	
Contract package name/discipline		Contract package reference	

What are the hazards?	Who might be harmed and how?	What are you already doing?	What further action(s) is/are necessary?	Action by who?	Action by when?
Heavy vehicle movement around the site.	General public, site operatives, other site visitors: ■ pedestrians being struck by delivery vehicle ■ lorry getting bogged down and becoming unstable.	Segregation of on-site vehicle and pedestrian routes. Vehicle only to reverse with the aid of a signaller. Site roads made up to a satisfactory standard; safe access to public highway. Everyone else kept out of the area during unloading.	Ongoing site inductions to include warning about keeping clear of vehicle unloading activities.	Site manager.	Before date of each delivery.

A 04

Risk assessments and method statements

Example of a qualitative risk assessment *continued*

What are the hazards?	Who might be harmed and how?	What are you already doing?	What further action(s) is/are necessary?	Action by who?	Action by when?
Mechanical lifting operations – unloading.	Delivery vehicle driver, others in the area: ■ contact with suspended load ■ falling/toppling load ■ lorry unstable during lifting.	Drivers must report to office on arrival for induction and checking of competence card – lorry loaders. Materials lay-down area ready for heavy, stacked materials. Other people kept out of the area.	Qualified slinger made available if requested by vehicle driver.	Site manager.	Before each delivery is unloaded.
Mud deposited on public roads.	General public – road users: ■ slippery road surfaces.	Vehicle wheel-wash installed; mandatory use when site is muddy.	Road sweeper is available if needed; mandatory use when site is muddy.	Site manager.	Immediately before vehicles leave site.

For further information refer to
www.hse.gov.uk/risk/fivesteps.htm

05

Leadership and worker engagement

What your employer should do for you	78
What you should do as a supervisor	79
Introduction	80
Communication	81
Good practice	82
The benefits of worker involvement	83
How is safety awareness achieved within a company?	83
Introduction to behavioural safety	86
What are the human factors?	86
The challenge	87
The solution	88
Communication with non-English speaking workers	89

Leadership and worker engagement

A05

	What your employer should do for you
1.	Consult with and involve all workers, provide training and match activities to capability.
2.	Attend to health and safety matters as a part of good management.
3.	Understand that a reputation can be lost by poor industrial relations.
4.	Recognise that accident rates are lower where workers are involved.
5.	Undertake training in communication skills and understand the importance of feedback.
6.	Appreciate that response in an agreed time is important.
7.	Understand that people's behaviour influences safety in the workplace.
8.	Involve the workforce in decision making, encourage suggestions and provide feedback.
9.	Encourage the workforce to be actively involved in enhancing safety and management systems.
10.	Design workplaces for the work activity.
11.	Provide information in a way or form that it is easily understood by anyone receiving it.
12.	Train staff in the recognition of workers with little or poor understanding of English.
13.	Provide opportunities for face-to-face discussions.
14.	Recognise workers who have reading, writing or hearing difficulties, or visual impairment.
15.	Ensure that workers get the right information at the right time.
16.	Provide supervisory staff with visual tools to aid safety-critical communication for those who have poor or no spoken understanding of English.

Leadership and worker engagement

What you should do as a supervisor

Checklist	Yes	No	N/A
1. Assist in consultation with and involving workers, provide training and match activities to capability.			
2. Pay attention to health and safety matters as a part of good management.			
3. Be aware that a reputation can be lost by poor industrial relations.			
4. Involve workers when planning safe systems of work.			
5. Participate in training in communication skills and understand the importance of feedback.			
6. Ensure that responses in an agreed time are important.			
7. Appreciate that people's behaviour influences safety in the workplace.			
8. Encourage the workforce in decision making, support suggestions and give feedback.			
9. Provide support to the workforce to be actively involved in enhancing safety and management systems.			
10. Ensure that workplaces are designed for the work activity.			
11. Explain information in a way or form that is easily understood by anyone receiving it.			
12. Recognise workers with little or poor understanding of English.			
13. Encourage opportunities for face-to-face discussions.			
14. Appreciate that workers who have reading, writing or hearing difficulties, or visual impairment, need help.			
15. Ensure that the work team get the right information at the right time.			
16. Use visual tools to aid safety-critical communication for those who have poor or no spoken understanding of English.			

A 05

Leadership and worker engagement

Introduction

An essential part of health and safety management is the two-way communication process between employers (or their representatives) and employees. There are legal duties on both parties with the intent that communication takes place in both directions.

> **Employers and their managers, supervisors, and so on, must talk to, and listen to, the people who work for and with them.**

Talking to, listening to and involving employees helps to:

- ☑ make the workplace healthier and safer
- ☑ improve motivation and productivity
- ☑ raise standards.

Good preparation helps to gain the commitment of the employees (or their representatives), so that they feel involved and enthusiastic about tackling health and safety together.

> **For further information refer to GE 700 *Construction site safety*, Chapter A09 Leadership and worker engagement.**

Helping to look after your business

Good health and safety management and a successful business are complementary. Your employer should already have in place ways of cutting down losses and waste reduction measures. Properly applied, these controls should also help you manage health and safety. You will want to help to do this so that there are well-trained people who are both healthy and safe.

If the company loses key people through inadequate relationships and poor health and safety performance, the business will be put at risk. The key is to get the workforce to recognise that managing health and safety is important and their top priority.

Helping to look after your reputation

The public and workers expect the HSE to take strong enforcement action. Failures can bring penalties of imprisonment or unlimited fines, as well as adverse publicity which will:

- ☑ put customers off doing business with your company
- ☑ prejudice their position on any pre-qualification or preferential supplier lists
- ☑ spread a bad reputation more quickly through the industry than a good performance.

Looking after your people

When staff are well protected, involved and well trained they will add value to the business because they:

- ☑ are better motivated
- ☑ take less sickness absence
- ☑ show greater loyalty.

To help in involving the workforce you need to:

- ☑ set them a good example
- ☑ train them well
- ☑ listen to their concerns
- ☑ encourage them to suggest solutions to problems
- ☑ provide feedback on how any concerns raised will be dealt with.

If raising a health and safety concern, it is essential that workers can raise it with someone on site who has the authority to take the issue forward as is appropriate.

The person raising the issue should be given assurance that their concern will be investigated and that feedback will be provided on the outcome.

The best way to protect the business, reputation and the workforce (and others) is to involve them.

- ✓ Talk to each other about issues.
- ✓ Listen to their concerns.
- ✓ Seek and share views and information.
- ✓ Discuss issues in good time.
- ✓ Consider what employees say before decisions are made.

Communication

The HSE's 'Fit3' campaign (Fit for work, fit for life, fit for tomorrow) resulted in the following findings.

- ✓ Accident rates are lower where employees have a say in health and safety matters.
- ✓ Employee involvement relates to a more positive health and safety culture.
- ✓ Stronger employee involvement makes for better control of common workplace risks.
- ✓ Employers can learn about risks through consultation.
- ✓ Employers with health and safety committees have less work-related injuries.

Understanding how we communicate

It is estimated that verbal communication only makes up around 10% of communication as a whole. Other aspects include such areas as sincerity, honesty, eye contact, body language, facial expression and timing – not forgetting that a picture tells 1,000 stories (that is, it is not so much what we say but how we say and present it).

Assertive communication

The most effective and healthiest form of communication is to use an assertive (confident) style. It's how we naturally express ourselves when our self-esteem is intact, giving us the confidence to communicate without games and manipulation. The style of communication will generally depend on your past experiences and lessons learnt to get the most effective results.

What should we be consulting our workforce about?

In general we must consult about:

- ✓ any changes that may have an effect on the workforce's health, safety and welfare
- ✓ arrangements for getting competent people to help the company meet what legally is required for health and safety obligations
- ✓ information on the likely risks in the workplace and the precautions that need to be taken
- ✓ the best way for information to be shared (consider language, literacy and learning disabilities)
- ✓ the planning of health and safety training
- ✓ health and safety consequences of introducing new technology.

Don't limit the scope of consultation to a pre-set list because there will be times when you should involve employees (and others) who are not on the list.

Leadership and worker engagement

Good practice

It is good practice to:

- ☑ provide feedback to explain the decisions and responses to issues
- ☑ involve employees in addressing work-related health issues, such as:
  - stress in the workplace
  - musculoskeletal disorders
  - sickness or injury
- ☑ commit to early involvement as a matter of routine
- ☑ agree to respond to issues within a certain timeframe
- ☑ solve problems jointly where employees participate as equals to resolve issues
- ☑ have a plan in place in case of inability to engage the workforce for reasons beyond immediate control.

Consultation does not always result in agreement but your company should be able to resolve differences of opinion by being open, explaining the reasons behind decisions and following agreed procedures for resolving problems.

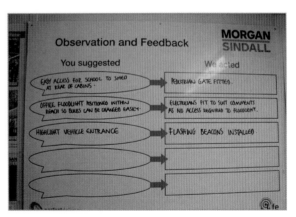

A 'you suggested – we acted' board helps workers see that concerns are being listened to and acted upon

Leadership and worker engagement

The benefits of worker involvement

Aside from their legal duty to consult, workplaces where employees are involved in taking decisions about health and safety are safer and healthier. Employees influence health and safety through their own actions. They are often the best people to understand the risks in their workplace. Engaging the workforce has proven to have major benefits in the workplace for all involved.

Talking, listening and co-operating with each other can help:

- ☑ identify joint solutions to problems
- ☑ develop a positive health and safety culture where risks are managed sensibly
- ☑ reduce accidents and ill health, plus their related costs to the business
- ☑ bring about improvements in overall efficiency, quality and productivity
- ☑ meet customer demands and maintain credibility
- ☑ comply with legal requirements.

A happy and healthy workforce is a hard working and safe workforce.

How is safety awareness achieved within a company?

The employer's role is to ensure that everyone, at all levels, is competent to do what is required of them and empower them to stop work and challenge any situation that they believe to be unsafe.

The supervisor's role is to support managers in their objective of achieving safety awareness with the objective of reducing incidents and accidents.

Everyone contributes to a safe workplace

Provide adequate information, instruction and training, for example when:

- ☑ a new site is started
- ☑ new people start on site
- ☑ new equipment is introduced
- ☑ a new phase of the job or major activity starts
- ☑ new safety procedures are required because of changes to work practices or routines
- ☑ at any other appropriate times (for example, daily briefings).

Leadership and worker engagement

Proactively manage day-to-day health and safety matters on site by:

- ✓ ensuring the company health and safety policy is adhered to
- ✓ ensuring that there are well-defined roles and responsibilities
- ✓ monitoring the day-to-day standards of health and safety on site
- ✓ ensuring that adequate training, information, training instruction and supervision are provided
- ✓ providing a means for employees to voice any concerns they have on health and safety issues
- ✓ acknowledging own limitations and seeking access to specialist health and safety advice where necessary.

Introduce safe systems of working by:

- ✓ implementing adequate written procedures of work activities and other relevant systems (for example, method statements and permits to work where appropriate)
- ✓ carrying out risk assessments, and implementing procedures required to control the problems identified by those assessments.

Introduce regular inspection schedules to include:

- ✓ audit inspections
- ✓ improvement plans.

Ensure that sound maintenance arrangements are carried out:

- ✓ using competent staff
- ✓ at appropriate intervals
- ✓ in the correct environment
- ✓ with adequate facilities being provided.

Develop and initiate plans for responding to:

- ✓ serious accidents or occurrences that require investigating or reporting
- ✓ trespassers or anyone causing damage on site
- ✓ site evacuation in the event of an emergency situation.

Engage (consult) your workforce on health and safety issues by:

- ✓ involving them in the decision-making process on risk reduction
- ✓ listening to their concerns
- ✓ taking positive action to reduce risks where their observations are valid
- ✓ providing feedback on how their concerns have been dealt with.

If health and safety matters are to be taken seriously by the workforce, management must make visible commitments to health and safety and lead by example.

Attention to health and safety issues must be seen as an integral part of the job and not just as a bolt on. Operatives should become accustomed to doing any kind of work with consideration to the health and safety of anyone who could be harmed, including themselves.

Safety awareness – worker toolkit.

The HSE has introduced a leadership and worker involvement toolkit. It is aimed particularly at small and medium sized businesses and is designed to help improve health and safety and bring additional benefits to the business performance and productivity.

The SLAM technique

The SLAM (Stop…Look…Assess…Manage) technique reminds workers to stop work if they think that their health and safety are at risk.

Why should you encourage workers to use SLAM?

By using the SLAM technique, workers will value the importance of health and safety and so create a healthy and safe site. By remembering SLAM, workers are more likely to stop work if a task appears unsafe or risky to their health, or stop their colleagues behaving in an unsafe or unhealthy way.

How to use the SLAM technique

The four stages of SLAM are shown below.

1. Stop the task and think. Look at each step and ask the following questions.

- ☑ Is this a new task?
- ☑ Has the task changed?
- ☑ When was the last time this task was carried out?
- ☑ Is everyone comfortable doing this task?
- ☑ If not, is there a need for training?

2. Look before, during and after completion of the task. Always:

- ☑ inspect the work area for potential hazards (such as unsecured ladders or untidiness)
- ☑ identify the hazards for each step of the job/task
- ☑ evaluate what to do about any hazards identified.

3. Assess if workers are equipped to perform that task safely. Check that everyone has the correct:

- ☑ knowledge
- ☑ skills
- ☑ training
- ☑ tools.

What else do they need to perform the task safely?

- ☑ Help? (Workers should be encouraged to ask for help.)
- ☑ More training? (Workers should not perform the task until they have been trained.)

4. Manage by taking appropriate action to eliminate or minimise any hazards on site by:

- ☑ ensuring the proper equipment is used and is well maintained
- ☑ thinking about the task just completed and asking:
  - what went well?
  - what did not go well?
  - did anything unexpected happen?
  - how can I be better prepared and plan for this in future?

What next?

- ☑ Share this information with workers and management.
- ☑ Use this information to encourage safe working practices.
- ☑ Use safety briefings and toolbox talks to teach your workers about the SLAM technique.

 For further information refer to www.hse.gov.uk/construction

There are other processes available, such as the STAR principle. *(Refer to Chapter A04 Risk assessments and method statements)*; others include SUSA, STOP, MAD, SAFE and Take 5 or Take 10.

Introduction to behavioural safety

Behavioural safety attempts to determine why people act the way they do in relation to work activities. It is based on a process of observation and feedback, and aims to identify, in advance, any difficulties in completing tasks safely.

It is proactive, trying to head off potential future problems rather than reacting to past accidents and mistakes.

Whilst the term *behavioural safety* is used throughout this section, the principles apply equally to preventing occupational health problems by influencing the behaviour of individuals and groups.

Why use behavioural safety?
Historically, improvements in health, safety and environmental performance have been achieved through improvements in engineering technology, that is non-human ways, and enhancing safety management systems. In many ways, this methodology has reached its peak performance and the related improvements in health and safety performance have begun to level off.

Future performance gains will only be achieved by taking more account of the way people interact in every aspect of the workplace; and through integrating and understanding the human element of risk management.

We all have our own perception of risk and safety based on our individual experiences and it is not easy to make direct comparisons between different views and opinions. Despite this, most people have a genuine desire to work safely through adopting best practice.

 The following HSE publications give clear guidance on managing safety and human factors:

- ✓ *Reducing error and influencing people* (HSG48)
- ✓ *Successful health and safety management* (HSG65).

These documents define *behavioural safety* as:

 organisational, job and individual factors which influence behaviour at work in a way that can positively affect health and safety.

To put it another way, the 'human factors'.

 For further information refer to GE 700 *Construction site safety,* Chapter A08 Behavioural safety and CITB's Achieving Behavioural Change (ABC) course.

What are the human factors?

The organisation: organisational factors, like workplace layouts, have the greatest influence on behaviour yet they are often overlooked during the design of work and investigation of accidents and incidents. There is a need to establish a positive health and safety culture.

Leadership and worker engagement

The job: tasks should be designed in accordance with ergonomic principles to take into account limitations and strengths in human performance (such as physical restrictions – matching the person to the task).

The individual: people bring personal attitudes, skills, habits and personalities to the job that can be strengths or weaknesses depending on the task. Generally, personalities cannot be changed but skills and attitudes may be changed and enhanced. Individuals are far more likely to respond positively to behavioural safety if they feel involved in the decision-making process; this is often referred to as worker engagement. They should be:

- ☑ asked for their views on how they feel that health and safety risks are being managed

- ☑ empowered to stop work if they feel that working conditions are not safe and encouraged to report the incident to someone in authority at site level

- ☑ provided with feedback on their suggestions, including how, if necessary, the issue is to be taken forward.

> For what reasons are workers tempted to carry out clearly unsafe practices?

Beliefs, expectations, attitude and behaviour

We all have our own beliefs (attitudes) that underlie how we think and define the way we act. If managers and supervisors are not committed or don't really believe that health and safety is a priority, a powerful negative message will be sent to employees. Low expectations and poor leadership from management can create negative attitudes from employees, resulting in poor methods of working that lead to poor health and safety performance.

Unsafe working at height

The challenge

Implementing a behavioural approach has always presented a challenge to industry because of the mind-set, culture, mistrust and constantly changing workforce and, until recently, attempts have often not been planned properly, or have been fragmented or reactive.

It is increasingly being recognised that integrating a systematic, proactive process within the organisation's arrangements can add significantly more value by addressing behavioural aspects of health and safety at the same time as optimising efficiency and productivity.

Furthermore, the HSE makes it clear that human factors must be taken fully into account when managing risk.

Significant improvements can be made to performance through an open communication and reporting process based on what is really happening and encouraging near-miss reporting. This will enable the better assessment of risk, bridges to be built, trust to be enhanced and the workforce to participate willingly as issues are resolved and solutions found.

The solution

A simple, fully integrated process that stimulates discussions on everything that is going on, whether or not they are related to safety, is progressed through an action plan and may include toolbox talks, weekly briefings, training and personal coaching. Once people see that these positive discussions lead to a positive gain, even without the difficult observation process that is generally unpopular, greater workforce involvement would occur.

Research has shown that when employers engage workers in discussions about health and safety, there is a reduction of accidents. (Source: HSE research report carried out by Glasgow Caledonian University.)

Managing change

Planning for the human side of change will make plans more likely to succeed. No single behavioural safety process fits into every company but some prescriptive processes may be a necessary step towards achieving open communication between the employer and the workforce.

Any change creates people issues (for example, there may be new leaders, changed roles and the need to develop new skills and capabilities). Employees may be uncertain and resistant because they do not see the need for change or feel that they will be disadvantaged by it.

Leading by example

The implementation of behavioural safety can pose particular problems with a fragmented and mobile workforce (such as that found in the construction industry).

To be successfully implemented on site, it is fundamental that the principles of behavioural safety are embedded within the organisation's culture and understood by the workforce and management from the beginning. It cannot just be thrown in as an initiative at a later stage.

The behaviour of supervisors and managers can directly affect the behaviour of operatives. The effect of failing to intervene in an unsafe situation is to condone that activity, practice or behaviour. This in turn sends a message to the operatives that the activity concerned is permitted and confuses the site teams. Therefore, intervention by managers and supervisors is critical in every case.

Communication

'Actions speak louder than words'. For trust to be built, an individual's behaviour and body language must reinforce what is being said –'walk the talk', as some people say.

Communication is at the heart of all that we do, both at work and in our own time. It is vital to give the person receiving information the time and space to be able to think and formulate a response. In communication, it is the quality, not the quantity, that matters.

 One method of enhancing any safe system of work is through frequent and open discussions.

The Health and Safety (Consultation with Employees) Regulations require employers to consult with their employees and make available such information, within the employer's knowledge, as is necessary to enable them to participate fully and effectively in the consultation.

Leadership and worker engagement

The benefits
Greater attention to improved working practices will bring about many benefits, including:

- ✓ reducing the potential for accidents
- ✓ creating a better system of work
- ✓ improved performance and greater awareness of issues and solutions
- ✓ a reduction in stress
- ✓ improved profitability.

Case studies
There is an undeniable link between behavioural safety (getting people to do what you want them to do) and actively engaging them in the decision-making process with regard to reducing site risks.

Examples of how engaging with the workforce has brought about tangible benefits can be found at www.hse.gov.uk/construction/engagement

Communication with non-English speaking workers

Good communication is essential for the management of health and safety on construction sites. The number of workers on UK sites, where English is not their first language, has increased over recent years. Some of these workers have excellent skills in spoken and written English, but there are others for whom understanding English is a problem. This can be a barrier to effective communication of health and safety information.

Several pieces of health and safety legislation, including the Health and Safety at Work Act 1974, require that employers pass comprehensible information, instruction and training to their employees. In the context of foreign workers, the word *comprehensible* can be taken to mean **provided in a format that can be** understood by the worker. Therefore if you employ non-English speaking workers on your site and fail to engage with them because of language barriers, not only are you likely to put them in danger, you will also be in breach of health and safety legislation.

One option for improving communication is for training materials to be represented in a pictorial form (images).

The effectiveness of images to overcome language barriers has been confirmed through research, which resulted in the development of a bank of images for this purpose.

CITB has developed a handy-sized, ring bound book titled *Safety critical communication – Toolbox talks* (GT 701). *(See the following pages for examples.)*

The images can be used to support site induction, toolbox talks and other training. They can also be used during spot checks to indicate the understanding of relevant issues. Furthermore, copies of the relevant images can be overlaid on site plans to indicate the location of welfare and first-aid facilities, fire-fighting equipment, the assembly point and so on.

Leadership and worker engagement

Each image is set out in the same format, comprising of four elements, as shown in the figure below.

Wear the correct PPE (protection) for cutting

1. Reference number
2. Phrase or sentence
3. Graphical symbol

4. Space for notes or translation

The second element is **a short phrase or sentence**. These have been developed, with help from language experts, to communicate, in plain English, the particular health and safety issue depicted by the graphical symbol.

Depending upon circumstances, a decision will have to be made as to whether it is more beneficial to reveal or hide the phrase or sentence during the assessment. Having sight of the phrase or sentence may have the potential to help workers with a limited understanding of written English to associate the written words with the message coming out of the graphical symbol.

The **graphical symbol** is the main element in each image. Most are self-explanatory. Some enable the person delivering the training to add site-specific information (for example, the postal address of the site or site telephone number).

As demonstrated in the figures below, it is recommended that some of the images are regarded as linked and shown sequentially to convey the correct meaning: if image 3.6 is shown to a worker who cannot understand the written phrase, it could be misinterpreted as 'the site is closed' or 'keep out'. However, by showing image 3.6 and 3.7 in sequence, a visual link is created and the chance of misinterpretation is reduced.

At the bottom of each image is a space for notes that may be used to add supplementary notes or, if employers wish, a basic translation into the workers' own language. Translation of short phrases like these is easier than translating several pages of text.

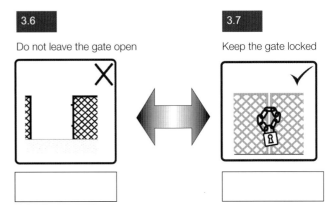

Do not leave the gate open

Keep the gate locked

Leadership and worker engagement

Delivering a safety critical communication toolbox talk

1. Find the hazard in the contents list.
2. Turn to the appropriate page (yellow image).
3. View the risk to health or safety on the opposing red page.
4. Fold out the red page to reveal the necessary control measures on the green page(s).
5. Deliver the toolbox talk.

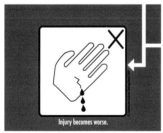

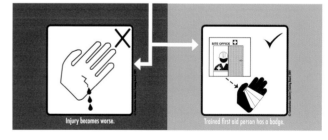

Leadership and worker engagement

Explanation of the elements on each coloured page

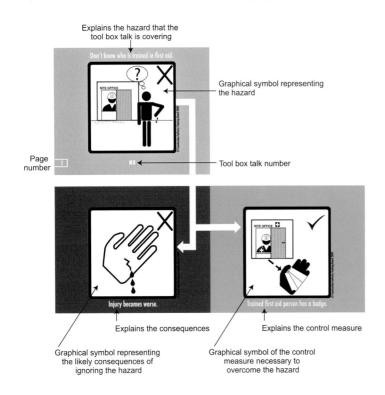

06

Statutory inspections, checks and monitoring

What your employer should do for you	94
What you should do as a supervisor	95
Introduction	96
Record keeping	96
Daily user and visual checks	96
Statutory or recommended inspections and examinations	97

Statutory inspections, checks and monitoring

What your employer should do for you

1. Provide plant and equipment with current examination/test certificates.
2. Ensure that examinations of plant are carried out in line with statutory requirements.
3. Provide formats for recording inspections and examinations and keep records.
4. Establish a system for marking unsuitable plant and equipment and putting it out of possible use (quarantine).
5. Implement a system of daily inspections and weekly recorded inspections.
6. Provide and record toolbox talks and monitor effectiveness.
7. Arrange for training of the workforce in the daily pre-use inspections and statutory examinations of plant and equipment.
8. Allocate responsibilities to the workforce for monitoring and control of activities.

Statutory inspections, checks and monitoring

What you should do as a supervisor

Checklist	Yes	No	N/A
1. Check that plant and equipment have current examination/test certificates.			
2. Make plant available for examinations in line with statutory requirements.			
3. Use forms for recording inspections and examinations and keeping records.			
4. Arrange for marking unsuitable plant and equipment and putting it out of possible use (quarantine).			
5. Carry out daily inspections and weekly recorded inspections.			
6. Present and record toolbox talks and report on effectiveness.			
7. Ensure your work team carry out daily/weekly checks and inspections.			
8. Request training of the workforce in the daily pre-use inspection of plant and equipment.			
9. Accept responsibilities for monitoring and control of activities of the work team.			

A06

Statutory inspections, checks and monitoring

Introduction

This chapter gives a brief outline of some of the requirements for the completion and use of various statutory and non-statutory forms, notices, signs and registers used within the building and construction industry, and the keeping of some records and other details.

Also included, some guidance on the types of daily and weekly inspection activities that supervisors should be carrying out and records that must be kept.

Record keeping

Record keeping is important as it is the means used to ensure that inspections and examinations are carried out.

Some are required by legislation and others by the employer as they are good practice and preventative and are vital in preventing accidents.

Records are generally split into two categories: those that are **proactive** (for example, records of inspections of scaffolding and excavations) and **reactive** (such as accident investigation).

Records of toolbox talks are also important as this demonstrates that (usually site-based) training is being provided.

Daily site briefings are often a good way of reviewing what has happened and looking at what has to be done and the associated hazards. This can be extended to point-of-work risk assessments *(refer to Chapter A04 Risk assessments and method statements* and *Chapter A05 Leadership and worker engagement).*

Daily user and visual checks

Daily user and visual checks form a vital part of good site management and reduce the risk of plant and equipment breakdown or failure, for example:

- ☑ engine/motor seizure due to lack of lubrication
- ☑ a low pressure tyre being damaged, or causing vehicle to overturn
- ☑ missing scaffold edge protection causing the potential of falling materials
- ☑ damaged power plugs, sockets and leads.

Scaffold inspection with inspection tags in place

Statutory inspections, checks and monitoring

Construction plant and equipment is exposed to harsh environments and they require effective maintenance regimes to avoid them developing defects. A programme of daily visual checks, regular inspections and servicing schedules should be established according to the manufacturer's instructions and the risks associated with the use of each vehicle.

Plant hire companies must provide information with all plant and equipment they supply to enable it to be used and maintained safely.

Contractual arrangements between user and hirer should set out who is responsible for maintenance and inspection during the hire period and these should be made clear to all parties.

Plant and vehicles should have a maintenance log to help manage and record maintenance operations. Employers should establish procedures designed to encourage supervisors and operators to report defects or problems, and ensure that problems with plant and vehicles are put right. Planned inspection and maintenance needs to follow manufacturer's instructions.

Statutory or recommended inspections and examinations

The following table sets out the recommended daily user checks, weekly inspections, statutory inspections and examinations.

For further information refer to GE 700 *Construction site safety*, Chapter A10 Inspections and audits and Chapter A11 Statutory forms, notices and registers.

Statutory inspections, checks and monitoring

Work activity plant item	Pre-use daily	Weekly record	Monthly record	Three monthly record	Six monthly record	12 monthly record	Form for statutory examination or report to comply with
		Statutory or recommended					
Excavations, cofferdams and caissons	✓ Inspect	✓ Inspect					Construction (Design and Management) Regulations
Plant and equipment (not electrical or for lifting)	✓ Inspect	✓ Inspect			✓ Examine	✓ Examine	Provision and Use of Work Equipment Regulations
Plant and equipment (electrical) including RCDs	✓ Inspect	✓ Inspect		✓ Examine			Construction (Design and Management) Regulations, Electricity at Work Regulations
Cranes and plant for lifting people, MEWPs, harness, lifting accessories and safety nets	✓ Inspect	✓ Inspect			✓ Examine	✓ Examine	Provision and Use of Work Equipment Regulations, Lifting Operations and Lifting Equipment Regulations
Cranes and plant used for lifting	✓ Inspect	✓ Inspect				✓ Examine	Lifting Operations and Lifting Equipment Regulations
Work at height, all scaffolds, working platforms, mobile towers, ladders and steps, etc.	✓ Inspect	✓ Inspect					Work at Height Regulations, Provision and Use of Work Equipment Regulations, Construction (Design and Management) Regulations
Fire-fighting appliances		✓ Inspect				✓ Examine	Construction (Design and Management) Regulations, Regulatory Reform Fire Safety Order
Site offices electrical equipment and installation		✓ Inspect				✓ Examine	Construction (Design and Management) Regulations, Electricity at Work Regulations, Workplace Health Safety and Welfare Regulations

07

Accident reporting and emergency procedures

What your employer should do for you	100
What you should do as a supervisor	101
What is an accident?	102
Common types of accident and incident	102
Accident prevention	104
Accident reporting	104
Reporting deaths, specified injuries and dangerous occurrences	105
Reporting occupational diseases	108
Keeping records	108
Post-accident investigation	109
First aid	110
Accidents and emergencies	112
Accident/incident reporting matrix	114
Suggested numbers of first-aid personnel to be available at all times people are at work	115

Accident reporting and emergency procedures

What your employer should do for you

1. Appoint a responsible person under RIDDOR.
2. Establish and maintain an accident reporting procedure.
3. Understand the definitions of specified injury and over three and seven-day injuries and inform employees.
4. Understand the term dangerous *occurrence* and train employees in recognition and understanding the importance of reporting.
5. Understand the meaning of the work-related illness, what they are and that they must be treated and reported.
6. Provide and maintain sufficient and suitable first-aid kits.
7. Assess and provide, where necessary, more extensive first-aid facilities where required.
8. Provide travelling first-aid kits for site vehicles and anyone who works in a remote location.
9. Assess needs for the provision of sufficient qualified first aiders and emergency first aiders as appropriate.
10. Arrange for sufficient appointed persons, where necessary, to provide temporary cover for first aiders and emergency first aiders.
11. Arrange for initial or refresher first-aid training, as necessary, to maintain adequate cover within the company.
12. Put in place suitable arrangements to quickly locate/contact first aiders and emergency first aiders.
13. Provide adequate first-aid signs, clearly displayed.
14. Maintain and make available accident books that comply with the Data Protection Act, to ensure confidentiality once completed.

Accident reporting and emergency procedures

What you should do as a supervisor

Checklist		Yes	No	N/A
1.	Accept duties as a responsible person under RIDDOR if requested.			
2.	Understand the accident reporting procedure and how to initiate a report.			
3.	Be aware of, and understand the definitions of specified injury and over three and seven-day injuries.			
4.	Understand the term *dangerous occurrence* and, if one occurs, know what actions to take.			
5.	Have an understanding of the types of work-related illness that must be reported.			
6.	Ensure first-aid kits are available, maintained and your work team understand where to obtain first-aid treatment.			
7.	Identify and report if you consider that more extensive first-aid facilities are required.			
8.	Ensure that travelling first-aid kits for site vehicles or remote locations are in place and replenished when necessary.			
9.	Inform your work team of the availability and location of first aiders or emergency first aiders and how to contact them.			
10.	Check that you have provided an appointed person in the event of temporary absence of the first aider.			
11.	Monitor the currency of certification and report any training requirements.			
12.	Maintain the display of adequate first-aid signs as appropriate.			
13.	Understand how to complete the accident book, its confidentiality once completed, and the reporting and investigation procedure.			

A07

Accident reporting and emergency procedures

What is an accident?

> An *accident* is an unplanned, unwanted, unscheduled event or occurrence which may result in injury to persons or damage to property.

It is noteworthy that:

- ☑ the injured person may not be an employee
- ☑ the property may not have been yours
- ☑ many accidents are easily preventable with a bit of forethought and monitoring.

Common types of accident and incident

Year after year, the same types of accident and incidents are repeated whilst carrying out construction activities. In many cases, taking simple precautionary measures would have prevented injuries and ill health.

Are you effectively managing the following?

Working at height – are risk assessments carried out and is the work properly planned, carried out and supervised by competent people?

Slips, trips and falls – is the site generally, and all access routes in particular, kept free from debris, materials and other tripping and slipping hazards?

Site plant operations – are all plant operators trained and authorised, and have effective measures been taken to segregate both plant that is operating and people on foot?

Manual handling – have manual handing activities been assessed and are work activities organised to minimise the need for strenuous manual handling?

The presence of **hazardous substances** – has the use (or presence) of all hazardous substances been assessed and are measures in place to eliminate or reduce exposure?

How bad is the accident situation in the industry?

For many years the construction industry has had a disproportionately high number of fatalities and non-fatal accidents, given the percentage of the total working population of the UK that work in the industry. It is quite possible that official accident statistics do not tell the whole story as anecdotal evidence suggests that as many as 40% of reportable accidents may not actually be reported. This is a criminal offence.

What causes accidents to happen at work?

There is no easy answer to this question. We can, however, identify some common causes.

- ☑ Poor communication.
  'I wasn't told how to do the job', 'I thought they knew that.'
- ☑ Lack of safety awareness.
  'I didn't realise that ...'
- ☑ Lack of concentration.
  'I've done this job so often I let my mind wander just for a second.'
- ☑ A lack of training/information/instruction.
  'The foreman just told me to get on with it.'
- ☑ Unfamiliarity.
  'But it's similar to other tools I've used so I never thought it would happen.'

Accident reporting and emergency procedures

- ☑ Poor decision making.
  'I needed a quick job done, so I suppose I cut a corner.'
- ☑ Poor attitudes.
  'Who needs instructions on a job like this?'
- ☑ Accidents only happen to others.
  'In all my years of experience, I've never had an accident yet.'

The Health and Safety Executive (HSE) has attributed many accidents to poor, or a total lack of, management and/or supervision. Whilst it may be the first reaction to blame the operative for the types of accident described, they are usually caused by a combination of circumstances and events. A competent employer, manager or supervisor should detect that a situation is becoming unsafe and intervene as necessary.

Employers, managers and supervisors have a responsibility for ensuring that operatives receive adequate instruction, training and supervision, and follow the specified safe system of working. Different groups may require differing levels of supervision.

A 07

Unsafe conditions

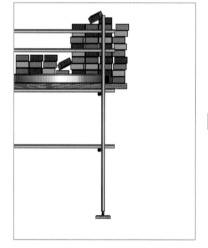

Something with the potential to cause harm

Near misses

An incident that nearly resulted in an injury or damage

Accidents

An incident that resulted in an injury or damage

Accident reporting and emergency procedures

Accident prevention

 Prevention is better than cure.

This is a saying that has never been more appropriate than in the business of achieving high standards of health and safety at work.

Putting these words into practice requires:

- [✓] an understanding of what can go wrong
- [✓] developing safe systems of work
- [✓] carrying out risk assessments and implementing the control measures identified
- [✓] the correct attitude by everyone concerned (employers and managers must engage with the workforce)
- [✓] ensuring that equipment and the working environment are safe
- [✓] a robust system enabling anyone to report hazardous situations and be certain that they will be put right
- [✓] learning from past incidents.

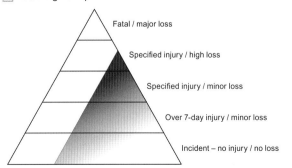

This illustration shows the inter-relationship between accidents with differing levels of severity. For every fatal accident there will be a greater number of specified injuries, even more over seven-day accidents and so on. The theory is that if the number of minor accidents can be significantly reduced, that is, the base of the triangle is shortened, it will have the effect of reducing the more serious events.

What do accidents cost?

To the victim	To employers
Life or death	Lost working time
Pain and suffering	Damaged equipment
Disability	Insurance costs
Financial loss, family conflict, dependence on others	Criminal prosecution or civil action
Reduced job expectations	Worry and stress
	Lost business, damaged reputation

Accident reporting

The Reporting of Injuries, Diseases and Dangerous Occurrences Regulations (Revised) (RIDDOR) place a legal duty on the responsible person to report certain events to the appropriate enforcing authority, either the HSE or the Local Authority (Environmental Health Officer). These regulations have been revised and came into force on 1 October 2013. Changes to these regulations have been included in the text on the next few pages.

Accident reporting and emergency procedures

In the case of the construction industry:

- ☑ the responsible person will be the employer in the case of an injury or other event affecting an employee or the person in control of the site where the affected person is self-employed or a member of the public

- ☑ the enforcing authority will usually be the HSE.

The responsible person must report any of the following events that happen in connection with work.

- ☑ Fatalities.
- ☑ Specified injuries *(see page 106)*.
- ☑ Any injury that results in an employee not being able to either come to work or carry out their normal work for more than seven days.
- ☑ An injury to a member of the public that results in the injured person being taken to hospital directly from the scene of the accident.
- ☑ Specified occupational diseases.
- ☑ Certain dangerous occurrences.

 Trainees are not specifically mentioned within these regulations, but other regulations require that non-employed trainees be regarded as employees for all health and safety purposes.

Reporting deaths, specified injuries and dangerous occurrences

The procedure for reporting deaths, specified injuries, dangerous occurrences and specified work related illnesses will usually be set out in detail in company safety policies and associated documentation.

The main requirements of the regulations are summarised on the following pages.

 Where an accident or occurrence has to be reported under RIDDOR, that report should be submitted online at www.hse.gov.uk/riddor

Tel: 0845 300 9923

All incidents can be reported online but a telephone service remains for reporting fatal and specified injuries **only** – call the Incident Contact Centre on 0845 300 9923 (opening hours Monday to Friday 8.30am to 5pm).

 Form F2508 can be completed online from www.hse.gov.uk/forms/incident/index.htm

The following must be **reported immediately** to the appropriate authority by the quickest practical method (usually by telephone) and reported on the approved form (F2508) within ten days:

- ☑ the death of any person as a result of an accident at work
- ☑ an accident to any person at work resulting in a specified injury or serious conditions specified in the regulations *(see list on p106)*

Accident reporting and emergency procedures

- [✓] any of the dangerous occurrences listed in the regulations *(see following list).*

 All fatal accidents must be reported to HM Coroner via the local police station. Police officers have a statutory duty to investigate fatalities in conjunction with HSE inspectors and the Crown Prosecution Service. The police investigation takes precedence in the event of a fatality. *(See also the earlier Corporate manslaughter section.)*

Exceptions

In general reports are not required for deaths and injuries that result from:

- [✓] road traffic accidents, unless the accident involved:
  - the loading or unloading of a vehicle
  - work alongside the road, such as construction or maintenance work
  - the escape of a substance being conveyed by the vehicle
- [✓] train accidents.

Specified injuries and serious conditions

Injuries and conditions that are classified as specified injuries for the purposes of RIDDOR include (as defined by the HSE):

- [✓] a fracture, other than to fingers, thumbs and toes
- [✓] amputation of an arm, hand, finger, thumb, leg, foot or toe
- [✓] permanent loss of sight or reduction of sight
- [✓] crush injuries leading to internal organ damage

- [✓] serious burns (covering more than 10% of the body, or damaging the eyes, respiratory system or other vital organs)
- [✓] scalpings (separation of skin from head) which require hospital treatment
- [✓] unconsciousness caused by head injury or asphyxia
- [✓] any other injury arising from working in an enclosed space that leads to hypothermia, heat-induced illness or requires resuscitation or admittance to hospital for more than 24 hours.

Reportable dangerous occurrences likely to occur in the building and construction industry

 Reportable and other dangerous occurrences where someone is nearly injured are called near misses. You must encourage your staff to REPORT NEAR MISSES to you so that you can learn from them and prevent a reccurrence of the situation.

Examples of dangerous occurrences that are reportable under RIDDOR are:

- [✓] collapse, overturning or failure of any load-bearing part of any lifting equipment, for example a winch, lift, hoist, crane, derrick, mobile-powered access platform, access cradle, window cleaning cradle, excavator, piling rig, forklift truck or lorry loader (for example, Hiab)
- [✓] explosion, bursting or collapse of any closed vessel, boiler and so on
- [✓] contact with, or arcing from, any overhead electric cable caused by any plant or equipment
- [✓] electrical short-circuit with fire or explosion

Accident reporting and emergency procedures

- ✓ collapse or partial collapse of any scaffold over 5 m in height or fall of any cradle and so on
- ✓ collapse of 5 tonnes or more of any building or structure, or any false work, or any wall or floor in any workplace
- ✓ uncontrolled release of any biological agents likely to cause severe human infection or illness
- ✓ any explosion, discharge or intentional fire which causes any injury to a person that requires first-aid or medical treatment
- ✓ the failure of equipment involved with pipeline works which could cause personal injury to any person, or which results in the pipeline being shut down for more than 24 hours
- ✓ malfunction of any breathing apparatus whilst in use or when being tested before use
- ✓ contact with, or arcing of, any overhead power line.

There are a total of 27 categories of dangerous occurrences that are relevant to most workplaces. For a full, detailed list refer to online guidance at www.hse.gov.uk/riddor

An easy guide chart is included at the end of this chapter.

'Over three-day' accidents

You must still keep a record of the accident if the worker has been incapacitated **for more than three consecutive days**, under the requirements of the Social Security (Claims and Payments) Regulations. But it is not reportable to the HSE until seven days after the injury accident.

'Over seven-day' accidents

The responsible person must report within 15 days, on the approved form (F2508), any workplace accident resulting in an injury that prevents a person from coming to work, or carrying out their normal type of work, for more than seven consecutive days.

In calculating the over seven-day period:

- ✓ the day that the accident occurred is not counted
- ✓ any rest days (for example weekends and bank holidays) that the injured person would have not been able to come to work had they been working days, are counted.

If an employee is injured on a Thursday and still off work through the same injury on the following Friday, the accident becomes reportable.

Delayed death

If an employer becomes aware of the death of an employee that:

- ✓ resulted from a previous reportable accident
- ✓ occurred within one year of the original accident

the employer must report the death to the enforcing authority within ten days of becoming aware of it, on an approved form (Form F2508) even though the original injury had been previously reported.

Accident reporting and emergency procedures

Reporting occupational diseases

The occurrence of certain occupational diseases, if resulting from particular work activities or are likely to be made worse, must be reported to the enforcing authorities by the quickest possible means. The form used for reporting occupational disease is F2508A.

The methods available for reporting occupational disease are the same as for reporting accidents and dangerous occurrences.

The occupational diseases (with possible sources) that are more common to the construction industry are:

- ☑ hand-arm vibration syndrome (using hand-held vibrating tools)
- ☑ carpal tunnel syndrome (similar to above)
- ☑ silicosis (from stone-cutting or working with masonry)
- ☑ occupational dermatitis (contact with any one of many types of substance that irritate the skin)
- ☑ occupational asthma (breathing in the fumes of any one of the many substances that irritate the respiratory tract (airways))
- ☑ severe cramp of the hand or forearm
- ☑ tendonitis or tenosynovitis of the hand or forearm
- ☑ any occupational cancer (such as asbestosis from refurbishment or demolition)
- ☑ any disease attributed to an occupational exposure to a biological agent
- ☑ illness resulting from exposure to a biological agent, such as legionellosis (from working on air conditioning systems) or leptospirosis (contracted while working in places likely to be infested with rats or other small mammals).

Keeping records

Company records of events that are reported under RIDDOR must be kept. No precise method is prescribed, but a photocopy of the approved form is acceptable, as are computer files and the transcript of reports made by telephone.

If an in-house accident form provides for the recording of the same details as on the approved form, it is acceptable.

The minimum particulars that must be kept are the:

- ☑ date and time of the accident or dangerous occurrence
- ☑ injured person's full name and occupation
- ☑ nature of injury.

In the event of an accident to a non-employee the minimum details required are:

- ☑ the injured person's full name and status (for example, passenger, customer, visitor or bystander)
- ☑ the nature of the injury
- ☑ the place where the accident or dangerous occurrence happened
- ☑ a brief description of the circumstances in which the accident or dangerous occurrence happened
- ☑ the date on which the event was reported to the enforcing authority
- ☑ the method by which the event was reported.

Accident reporting and emergency procedures

Accident book

Employers are required under the Social Security (Claims and Payments) Regulations to keep an accident book readily available, into which details must be entered of every accident causing personal injury to any employee. Wherever possible, each entry should be made by the injured employee. Where this is not possible, entries may be made by anyone acting on their behalf.

> Making an entry in the accident book does not meet or replace the employer's obligation to report specific accidents and dangerous occurrences to the HSE under RIDDOR.

Employers may use an official Accident Book BL510 or may develop an in-house method of recording accidents (paper or electronic), providing it enables the recording of all the relevant details.

Under the Data Protection Act, accident details are confidential between the employer and injured person once a record has been completed. It is essential therefore that irrespective of what method of accident recording is used, it prevents anyone who is making a subsequent entry from seeing previous accident records. This is achieved if using Accident Book BL510, as it has perforated pages, which can be detached and stored in privacy.

All accident books or other forms of accident record must be kept for three years from the date of the last entry. Health surveillance records must be kept for 40 years.

The above arrangements may provide essential information in compensation claims and other actions. They do not, in any way, relieve the employer of the responsibility to provide the HSE or Local Authority with any information the law may require.

Failure to report an accident is a criminal offence and lack of company evidence of an incident having occurred will be harder for a company to defend if an injured person claims for compensation at a later date.

> All accidents should be investigated if lessons are to be learned from these experiences to ensure that control measures can be put into place so that they do not happen again.

If there are trade union-appointed safety representatives on site, they can assist in the investigation. Inspectors from the HSE will investigate the most serious accidents.

Post-accident investigation

Following an accident, attend to the need of any injured person first (as outlined later in this chapter) and then try to establish the events leading up to the accident.

- ✓ Note anything that appears significant, including making sketches, taking photographs and so on.
- ✓ Establish exactly what processes were being carried out.
- ✓ Find out whether a method statement and/or permit to work was in force and, if so, whether they were being complied with.
- ✓ Check whether the safe system of work, as determined by the risk assessment, was being followed.
- ✓ Find out who was operating any equipment that was involved in the accident and whether they were competent to do so.
- ✓ Find out if the injured person was authorised to be doing what they were doing or be where they were.

Accident reporting and emergency procedures

In many cases it will be necessary at some stage to:

- [✓] interview the injured person (if possible)
- [✓] question the person in charge of the process or project
- [✓] identify and interview witnesses.

It is most important to carry out an investigation into any accident as soon as possible after it has happened whilst events are still clear in the minds of the injured person(s) and any witnesses.

For further information refer to GE 700 *Construction site safety*, Chapter A12 Accident prevention and control.

First aid

The principles of first aid are:

- [✓] where a person requires assistance from a doctor or nurse, to preserve life and render other assistance until such help arrives
- [✓] to treat minor injuries for which a doctor or nurse is not required.

The revised Health and Safety (First Aid) Regulations came into force on 1 October 2013. The changes have been reflected in the following text.

Provision of first-aid equipment and staff

The Health and Safety (First Aid) Regulations, together with the Approved Code of Practice L74, require employers and the self-employed to assess the need for both first-aid equipment and qualified staff based upon a first aid needs assessment. The assessor should consider:

- [✓] the nature of the work carried out
- [✓] the number of employees and how they may be dispersed
- [✓] the remoteness of the site from any emergency services
- [✓] the needs of travelling or lone workers
- [✓] whether any arrangements have been made for the sharing of first-aid facilities with other employers
- [✓] the need for holiday and shift cover
- [✓] the history of accidents in the organisation.

First-aid staff

The HSE has published guidance to help you understand and comply with the regulations, and offers practical advice on what you need to do. The guidance includes details of an optional four-layer framework for first-aid provision. The layers are:

- [✓] appointed person (AP)
- [✓] emergency first aid at work (EFAW)

Accident reporting and emergency procedures

- [x] first aid at work (FAW)
- [x] additional training.

Based upon the findings of the assessment, the employer must:

- [x] appoint a sufficient number of suitable and trained first aiders and/or emergency first aiders to render first aid to employees and others, and/or
- [x] appoint persons who, in the absence of a trained first aider, will be capable of taking charge in an emergency, calling an ambulance and looking after first-aid equipment. They are not allowed to give first aid. Emergency first aiders undergo shorter training and are more restricted in what they can do.

 The HSE has published a guide for the suggested number of first aid personnel that should be available at all times. This is reproduced at the end of this section.

The figures in the guidance are a guide only and do not allow for any special circumstances (such as high risk activities, shift cover or groups working in remote locations). Even those who are not trained in first aid but know how to carry out resuscitation may help to save a life.

The training of employees to become first aiders or emergency first aiders may only be carried out by organisations that have the expected skills, qualifications and competence.

A certificate is issued, which is valid for three years, after which time a re-certification course is required. Anyone whose certificate has expired must undertake a full course of training to be re-established as a first aider.

 The HSE has developed a *First aid at work* assessment tool. It is designed to help employers determine the number and type of first-aid personnel to provide in their workplace.

It can be found at www.hse.gov.uk/firstaid/assessmenttool.htm

The HSE strongly recommends that first aiders and emergency first aiders undergo annual refresher training during their three-year certification period. Although not mandatory, this will help qualified people maintain their basic skill and keep up to date with changes to first-aid procedures.

 First aiders have the potential to save lives.

First-aid equipment

Based upon the findings of the needs assessment, the employer must:

- [x] provide and keep stocked suitable first-aid boxes and other appropriate equipment (such as eyewash stations and burns kit), as determined by the assessment, in places that can easily be accessed by all employees
- [x] display notices giving the identity of first aiders and the location of first-aid equipment.

The very minimum that would be expected on any site is a small basic first-aid kit. On larger sites several kits may be necessary at dispersed locations, and, depending on the results of the assessment, a small, travelling first-aid kit may be requested for those operatives who work alone or in remote locations. A travelling first-aid kit should be located in each company vehicle.

Accident reporting and emergency procedures

In situations where specific hazards exist, it will be necessary to consider providing more specialised first-aid equipment (such as rescue equipment and emergency showers).

Accidents and emergencies

Immediate action

If you are first on the scene where someone has suffered a serious accident:

- ☑ assess the situation to ensure there is no danger to yourself
- ☑ if it is necessary and possible, remove or isolate the source of danger
- ☑ only move the casualty if it is absolutely necessary
- ☑ unless absolutely necessary stay with the casualty and send for qualified medical help; initially this may be a first aider
- ☑ confirm that someone has called the emergency services
- ☑ comfort the casualty and keep them warm until help arrives
- ☑ use other people to isolate the danger area as far as is necessary
- ☑ as far as possible, arrange for the area to be kept clear of other people (bystanders) unless they are actually helping.

The company health and safety policy (arrangements) must explain in detail the procedure to be followed by an employee who is the first on the scene at an accident.

Only move a casualty to save a life or to prevent further injury.

After the casualty has been removed

Preserve any evidence as to why the accident occurred:

- ☑ arrange for barriers to be erected and warning signs to be displayed if some hazards still exist
- ☑ report the accident to the HSE and police if appropriate
- ☑ investigate what happened and why
- ☑ learn from the event and amend procedures if necessary.

Your employer should:

- ☑ communicate your company's first-aid arrangements to those who need to know
- ☑ be sure that they have taken in the information
- ☑ ensure that first-aid arrangements are updated as necessary, and are appropriate
- ☑ ensure that everyone knows what to do in an emergency.

Finally, as a supervisor:

- ☑ you have a duty to communicate your company's first-aid arrangements to those who need to know
- ☑ be sure that they have taken in the information and know what to do
- ☑ check that the first-aid arrangements are in place and report any shortfalls.

Does everyone know what to do in an emergency?

For further information refer to GE 700 *Construction site safety,* **Chapter B05 First aid.**

Accident reporting and emergency procedures

Emergency and first-aid procedures

Details of the site first aiders	Name		Telephone / location	
	Name		Telephone / location	
Location of first aid box(es)				
Site address, including postcode				
Contact details	Ambulance			
	Fire service			
	Other			
Location / address of nearest hospital, including postcode				
	Directions (see map)			
Where to record treatment given and follow up actions				

A 07

Accident reporting and emergency procedures

Accident/incident reporting matrix

Type of incident	Enter in firm's records and investigate	Put in accident book	Send Form 2508 to enforcing authority*	Phone enforcing authority and send Form 2508*	Send Form 2508A to the enforcing authority*
Any incident or near miss	✓				
Every injury**	✓	✓			
Over three days' time lost	✓	✓			
Over seven days' time lost ***	✓	✓	✓		
Specified injury or fatal	✓	✓	✓	✓	
Dangerous occurrence	✓		✓	✓	
Specified ill health*	✓				✓

*Refer to the Accident reporting section of this chapter for the different methods of submitting accident reports to the enforcing authority.
**Note employers and others with responsibilities under RIDDOR must still keep a record of all over three-day injuries. An accident book record will be enough.
***Note the deadline by which the over seven-day injury must be reported is 15 days from the day of the accident.

Accident reporting and emergency procedures

Suggested numbers of first-aid personnel to be available at all times people are at work

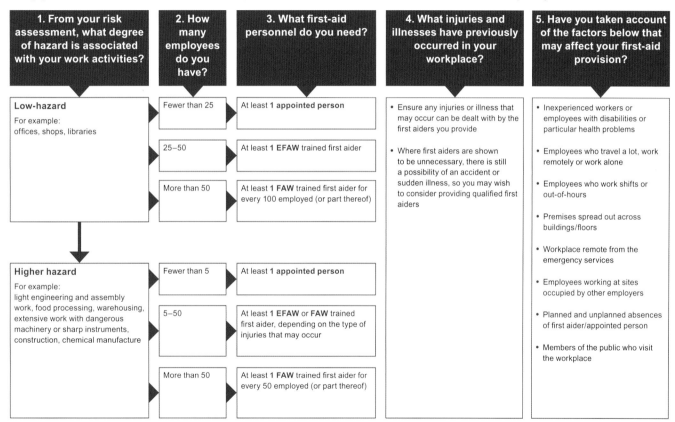

Source: HSE First aid at work guidance on regulations Appendix 3 (L74).

Accident reporting and emergency procedures

A07

08
Health and welfare

What your employer should do for you	118
What you should do as a supervisor	119
Introduction	120
Occupational dermatitis	120
Radiation	122
Sun safety	123
Leptospirosis (Weil's disease)	123
Drugs and alcohol	124
Health surveillance	126
Welfare facilities	127
Food safety	129

Health and welfare

What your employer should do for you

1. Provide toilets that are suitably ventilated and lit.
2. Provide washing facilities (including showers if necessary), which are equipped with hot (or warm) and cold water, soap and towels or a means of drying.
3. Ensure that there are rest areas (including facilities for a pregnant female or a nursing mother to lie down), as appropriate.
4. Provide drying and/or changing rooms with secure lockers and where necessary, separate facilities for men and women.
5. Ensure that the welfare facilities are clean, kept in good order and properly maintained.
6. Provide a supply of fresh drinking water, complete with suitable cups (unless from a water fountain or similar).
7. Ensure that there is a means of preparing food.
8. Make arrangements to enable food to be eaten in reasonable comfort (an adequate number of tables and chairs with backs).
9. Provide a means of boiling water.
10. Ensure controls are in place to protect people from radiation (for example, ultraviolet light, infrared sources from lasers and welding).
11. Provide health information and protective measures to people working in the sun.
12. Ensure adequate personal protective equipment (PPE) and washing facilities with good housekeeping controls are in place to minimise the risk of leptospirosis (Weil's disease).
13. Implement a drugs and alcohol policy, including prescription drugs awareness and a monitoring process for the workforce.
14. Monitor and implement, if appropriate, the need for health surveillance for potentially hazardous work activities.

Health and welfare

What you should do as a supervisor

Checklist	Yes	No	N/A
1. Ensure toilets are in good working order and kept clean.			
2. Check washing facilities, which are equipped with hot (or warm) and cold water, soap and towels or a means of drying, are known to the workers and report any defects.			
3. Ensure that rest areas are properly maintained and waste is removed regularly.			
4. Check that drying and/or changing rooms are kept clean and tidy and free from rubbish.			
5. Arrange for welfare facilities to be clean, kept in good order and properly maintained.			
6. Check that drinking water is available and cups (where appropriate) are available.			
7. Inspect the area for preparing food to ensure its suitability and report defects to your manager.			
8. Ensure that food can be eaten in reasonable comfort.			
9. Check the means of boiling water are safe.			
10. Assist in ensuring that controls are in place to protect people from radiation (for example, ultraviolet light, infrared sources from lasers and welding).			
11. Ensure that health information and protective measures are provided and explained to people working in the sun.			
12. Check to ensure adequate PPE and washing facilities with good housekeeping controls are in place to minimise the risk of leptospirosis (Weil's disease) and provide information to people affected.			
13. Report instances where the implementation of a drugs and alcohol policy may be necessary, including monitoring people on prescription drugs.			
14. Recognise and report on situations where you think that there is a need for health surveillance for potentially hazardous work activities.			

B 08

Health and welfare

Introduction

Occupational ill health is the term used with regard to health issues that arise out of work activities. This chapter covers the common occupational health topics associated with construction industry activities, including smoking restrictions. Even those employers that attempt to ensure the safety of their employees will often overlook the threats to employees' occupational health.

A reason for this might be that, in some cases, an occupational ill health problem may not be so obvious as the result of an accident (for example, a broken bone). Therefore, the potential seriousness of the situation may not be immediately apparent. Some forms of occupational ill health may take many months or years to develop.

It is a fact that at any one time there are more people off work as a result of occupational health issues than there are as a result of accidents.

For further information refer to Chapter B10 Dust and fumes (Respiratory risks).

Manual handling, which is also an occupational health issue, is covered separately in Chapter B13.

Occupational dermatitis

Dermatitis is a major cause of absenteeism, not only in the building and construction industry but across the whole spectrum of industry, and accounts for over half of all working days lost through industrial disease. It is a serious skin condition that is caused by irritants contained in many industrial materials. It cannot be passed on from the sufferer to other people. There are two general types, which are outlined below.

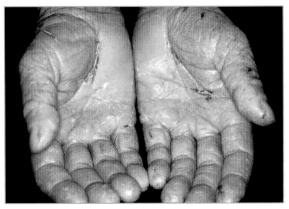

Dermatitis showing crusting and thickening of skin

Irritant dermatitis

This is usually caused either by the skin coming into contact with an irritant substance, which is usually a chemical or dust. Repeated exposure to extreme heat or cold can damage the skin, which makes it more likely that irritant dermatitis will occur. Anyone may be affected. The length of exposure, together with the strength of the irritant substance, will affect the seriousness of the complaint. Most cases of dermatitis are of this type.

Health and welfare

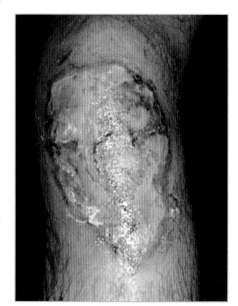

Irritant contact dermatitis 'pizza knee' from a cement burn

Personal hygiene is particularly important when working with materials that may be irritants, as resistance to an irritant varies with the type of skin.

Pores, ducts and hair follicles in the skin may admit irritants to the sensitive inner skin layer and, therefore, washing thoroughly to remove the dirt and grime with soap and water is an essential preventative measure. It is equally important that clothing is kept clean. Oil-stained and dirty overalls are a well known cause of skin problems around the groin.

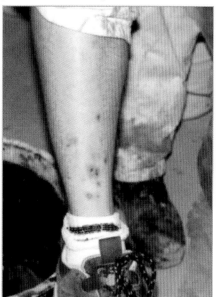

Allergic contact dermatitis of the leg

Sensitive (or allergic) dermatitis

Some people develop an allergic reaction to a specific substance. This reaction may follow after weeks, months or even years of use or exposure to a substance without any ill effects. However, once sensitive dermatitis has occurred, any future exposure to the substance will again produce an adverse reaction.

Reaction of the skin to irritants varies from one individual to another. The reaction may be only a mild redness, or it can develop into swelling, blisters and septic ulcers that are both unsightly and painful.

Health and welfare

Controlling the risk

Employers should identify the:

- ☑ substances used by employees that have the potential to cause dermatitis
- ☑ individual employees who are at risk
- ☑ work activities bringing the substance and employee together.

Control measures must be introduced to minimise the chance of skin contact with the substance and therefore the risk of dermatitis. While the risk from products that are purchased are easy to identify from information on the packaging, more effort will be required to determine the risks from dust (such as hardwood dust) that is produced by a work process.

Screening for the early signs of dermatitis can be carried out by simply conducting periodic hand/forearm checks of at-risk individuals by an employee who has been trained to recognise the symptoms. Where dermatitis is diagnosed, the sufferer must be advised to seek medical advice at the earliest opportunity.

Control measures

- ☑ Identify irritant substances.
- ☑ Substitute with a less harmful substance.
- ☑ Limit contact with harmful substances, for example:
  - dust extraction fitted to tools and machines
  - workplace ventilation
  - using the appropriate PPE (gloves, and so on).
- ☑ Frequent cleaning of protective equipment.
- ☑ Immediate washing of hands or other areas of affected skin.
- ☑ Use of reconditioning creams or barrier creams.
- ☑ Prompt reporting of skin irritation.
- ☑ Seeking medical advice.

Occupational dermatitis is a notifiable disease under the Reporting of Injuries, Diseases and Dangerous Occurrences Regulations. If in any doubt, contact your local Health and Safety Executive (HSE) office or the HSE Employment Medical Advisory Service (EMAS).

 For further information refer to Chapter B12 Hazardous substances and Chapter B09 Personal protective equipment.

Radiation

An increasing range of equipment that uses radiation sources as a measurement or detection device is now being introduced into industry.

Radiation that can cause concern includes ultraviolet light and infrared sources, also:

- ☑ microwaves
- ☑ lasers
- ☑ some welding processes.

Exposure to radiation may lead to:

- ☑ burns (including sunburn)
- ☑ eye damage (cataracts and arc-eye)
- ☑ cancers (skin cancer and leukaemia).

Health and welfare

Exposure may quickly reach danger levels in some cases. In others, the problems may develop over long periods.

Where a radiation source that has the potential to cause harm is being used, measures must be taken to keep those not involved out of the danger area.

If you are working near or notice the warning sign shown here, seek the advice of the person in charge on site.

 For further information refer to GE 700 *Construction site safety*, Chapter B05 First aid, Chapter B06 Personal protective equipment and Chapter B08 Skin protection.

Sun safety

Even a tan that has been built up gradually can be harmful to health. A tan is a sign that the skin has been potentially damaged.

The damage is caused by the ultraviolet rays in sunlight. People whose jobs keep them outdoors for long periods of time (such as construction workers) may, if their skin is unprotected, get more sun on their skin than is healthy for them. They will then be at a greater risk of developing skin cancer.

The dangers

In the short term, excess exposure of unprotected skin to the sun causes sunburn, which can blister the skin and make it peel. Even a mild reddening of the skin is a sign of skin damage. In the long term, too much sun will speed up the ageing of skin, making it leathery, mottled and wrinkled. However, the most serious issue is an increased chance of developing skin cancer. Cases of skin cancer have more than quadrupled over the last 30 years and more than two young adults (aged 15 to 34) are now diagnosed with skin cancer every day in the UK.

 Some medicines and contact with some chemicals used at work (such as bitumen products) can make the skin more sensitive to sunlight.

The risks

Some people are more at risk than others. People with pale skin are most at risk. Workers should take particular care if they have:

- ☑ fair or freckled skin that does not tan, or goes red or burns before it tans
- ☑ red or fair hair and light coloured eyes
- ☑ a large number of moles.

Leptospirosis (Weil's disease)

Leptospirosis (Weil's disease) can be caught from coming into contact with the urine of rats, voles, mice and other small animals. It can be fatal if not diagnosed and treated quickly.

The disease is usually associated with working in sewers, on sewage plants, on old farm buildings and on sites close to rivers or canal banks, but can be caught anywhere that the small mammals that

Health and welfare

carry the disease exist (such as any wet site). A simple course of action is to discourage the presence of rats by making sure that food waste is properly disposed of.

It is advisable that any workers who are particularly at risk be issued with a 'Leptospirosis risk' card or neck-tag so that if they become ill, the doctor's attention can be drawn to the possibility of leptospirosis, since the early symptoms of the disease are very similar to, and may be mistaken for, influenza.

The common routes of entry into the body are:

- ☑ unprotected cuts and grazes on the skin
- ☑ swallowing water infected with the bacteria
- ☑ absorbing water infected with the bacteria through the lining of the nostrils.

Anyone who has broken skin that could come into contact with anything that is contaminated should not be exposed to the risk. Preferably, someone else will be able to do the job or, as a very minimum, cuts and grazes will be totally covered by a secure waterproof dressing.

Risk management measures should ensure that no-one is put at risk by falling into water. The provision and use of adequate washing and, if necessary showering facilities, should further lessen the chance of anyone catching the disease.

Anyone who falls into, or is otherwise immersed in, water that is likely to be contaminated should promptly seek medical advice, particularly if any water was swallowed.

A less serious, although still unpleasant form of leptospirosis can be caught from the urine of infected cattle. The routes of entry into the body are through the eyes, nose, mouth and broken skin.
A safe system of work will prevent exposure, with PPE being used as a last resort. Construction work on farms and livestock markets (for example, where workers might get into the close proximity of cattle), presents the potential for those workers to catch this strain of leptospirosis.

Leptospirosis is a notifiable disease under the Reporting of Injuries, Diseases and Dangerous Occurrences Regulations (Revised) (RIDDOR). *(For more information refer to Chapter A07 Accident reporting and emergency procedures.)* If in any doubt, contact your local HSE office.

Drugs and alcohol

The use of drugs or alcohol gives rise to the risk of under-performance and the possibility that an employee will compromise their own health and safety and that of other people, including members of the public.

Alcohol

The effects of alcohol in the body can remain for long periods. It is generally accepted that alcohol dissipates in a healthy body at around half a pint of beer per hour. So, if you drink one pint of beer, your body takes about two hours to break it down. One pint of strong lager is equivalent to three units, so it will take longer. However, this time can vary depending on the factors such as your size, weight, age, when you last ate and any medication you are taking. Everyone should be aware of the morning after effects. Intakes of coffee and eating stodgy food will not lessen these effects.

Health and welfare

 Alcohol affects sensory perception and reaction times.

There are people at work that are never free of the effects of alcohol and, because of this, they are a constant source of danger to themselves or anyone else working with them.

 The National Health Service recommends a maximum intake of:

- ☑ 3-4 units per day for men
- ☑ 2-3 units per day for women

where one unit of alcohol is roughly equal to one small glass of wine, a single measure of spirits or half a pint of normal strength beer.

 More than 1 in 10 deaths of people in their 40s are from liver disease. Most are from alcoholic liver disease.

Prescription drugs

In addition to the dangers that can be caused on site by the use of illegal drugs, some of the drugs prescribed by doctors may have unwanted side effects. On every prescription there is a label giving details of the correct dosage to be taken and at what intervals. This dosage must be strictly adhered to, as taking more than directed may have adverse effects, particularly in the case of painkilling drugs and antihistamines.

Some direction labels may also give a warning, for example:

> May cause drowsiness. If affected do not drive.
> Do not operate machinery.
> Do not take with alcohol.
> Avoid alcoholic drinks while taking this medication.

Such warnings should not be ignored – they are there for the guidance and advice of the person for whom the drugs are prescribed and should be followed.

Over-the-counter medicines

Some medication that can be bought from retail outlets can also cause side effects that may lead to the person taking it being unfit for work on a construction site. (For example, some hay fever medicines cause drowsiness and therefore loss of concentration.)

Always read the instructions for the correct dosage and any information on the possible side effects.

Identifying alcohol and drug problems

Employers and managers should be aware of the various characteristics that may indicate a problem exists. These include:

- ☑ absenteeism
- ☑ poor time-keeping
- ☑ high accident level
- ☑ poor work performance
- ☑ mood swings
- ☑ misconduct
- ☑ theft, to feed personal habits.

B 08

Health and welfare

Managers and supervisors should be trained in recognising and responding to the early stages of an alcohol or drugs problem amongst the workforce.

An employee may come forward voluntarily and seek help. In other instances, the problem may be identified by a co-worker or by a supervisor. There is a need to ensure that employees who seek help will not be disadvantaged or punished. Confidentiality for employees undergoing any treatment or rehabilitation must be guaranteed.

Addressing alcohol and drugs at work

Consultation between a company and its employees, either directly or through their representatives, is essential if alcohol and drug problems are to be addressed effectively. Any such action must be supported by a drug and alcohol misuse policy, which all workers are made aware of, either as part of their induction into the company or prior to any such policy being introduced. In larger organisations, drug and alcohol misuse could be a standing agenda item for each meeting of the health and safety committee, even if at most meetings there is nothing to report.

For further information refer to GE 700 *Construction site safety*, Chapter B04 Drugs and alcohol.

Health surveillance

The exposure of employees to some occupational health hazards will require that those employees are provided with appropriate health surveillance. The purpose is to detect the early signs of an occupational health problem so that appropriate measures can be put in place. Should health surveillance detect any signs of ill health, it is an indication that existing control measures are not effective.

The extent of health surveillance will depend upon the nature of the potential health problem and how far it has progressed.

At one end of the scale, a suitably trained person carrying out periodic skin checks of anyone exposed to an irritant is an example of simple health surveillance that can be carried out on site.

An example at the other end of the scale is audiometry (hearing) checks carried out by a qualified technician using specialist equipment, at a hospital or health screening unit.

Employers and managers should ensure that employees seek qualified medical advice if on-site health surveillance or discussion with the employee indicates that there could be an occupational health problem that needs medical attention.

Generally for smaller construction companies, the need for health surveillance may result from exposure of employees to:

- ☑ noise
- ☑ vibration
- ☑ some hazardous substances (such as solvents and dust)
- ☑ asbestos
- ☑ lead.

Examples of the occupational health hazards that may create a need for health surveillance are exposure to:

- ☑ substances that have the potential to cause dermatitis or other skin conditions
- ☑ substances that have the potential to cause respiratory illness, including asbestos

Health and welfare

- ✓ levels of noise that have the potential to cause noise-induced hearing loss
- ✓ levels of hand-arm vibration that have the potential to result in occupational illnesses (such as vibration white finger)
- ✓ levels of whole body vibration resulting in back problems.

Welfare facilities

The thrust of the legislation relevant to welfare on site is broadly divided between the requirement for adequate rest and personal hygiene facilities and the requirements relating to food safety.

The Construction (Design and Management) Regulations place a legal duty on:

- ✓ the person in control of a site to provide suitable and sufficient welfare facilities for those who work on the site, so far as is reasonably practicable
- ✓ employers and the self-employed to ensure that suitable and sufficient welfare facilities are provided for their employees or anyone else working under their control.

The employer must ensure that welfare facilities are provided for employees, but they do not necessarily have to provide them personally. This allows for the sharing of welfare facilities where there is more than one contractor on site.

It should be noted that the duty to provide welfare facilities is placed upon the person in control of the site and not the employer, who will not be the same person in some cases. Contractors arriving to work on another contractor's site should fully expect to find an adequate range of welfare facilities, except where the person in control of the site has deemed it not reasonably practicable to provide them.

The principle of 'so far as is reasonably practicable' allows the person in control of a site to base a decision upon whether or not to provide welfare facilities by balancing the expense of providing them in terms of cost, time and inconvenience against the risks to the health and safety of those on site if they are not provided.

There is no question that appropriate welfare facilities should be provided on the vast majority of fixed construction sites. However, where a construction site is mobile (for example, a two-man gang carrying out minor road repairs), it might justifiably be deemed not reasonably practicable to provide the full range of welfare facilities at their various remote locations, but they must be available at nearby locations that have been previously identified.

General welfare requirements

The regulations require that:

- ✓ suitable and sufficient sanitary conveniences (toilets) must be provided or made available at readily accessible places
- ✓ suitable and sufficient washing facilities, including showers if required, must be provided or made available at readily accessible places
- ✓ an adequate supply of drinking water, conspicuously marked with the appropriate sign, should be provided or made available at readily accessible places
- ✓ suitable facilities must be provided for the accommodation of clothing not worn at work and for clothing worn at work but not taken home (such facilities shall include provisions for drying clothing when it gets wet)
- ✓ facilities must be provided to enable people to change clothing where a person has to wear special clothing for special work and cannot change elsewhere (the facilities shall be separate for men and women, where necessary, for reasons of health and privacy)

Health and welfare

- ☑ suitable and sufficient facilities for rest (such as a site canteen) must be made available at readily accessible places, including provisions that:
  - sufficient tables and chairs (with backs) are provided
  - there are suitable arrangements to protect non-smokers from discomfort caused by tobacco smoke *(see 'Smoking restrictions' in Chapter C15 Fire prevention and control)*
  - where necessary, facilities are suitable for pregnant women and nursing mothers
  - a means for boiling water is provided and there are suitable arrangements to ensure that meals can be prepared and eaten.

No matter how basic or extensive the welfare facilities, they must be properly maintained, cleaned and well ventilated

Welfare facilities in general should be kept clean, adequately lit, ventilated as necessary and kept in a good state of repair.

Washing facilities must include:

- ☑ a supply of hot (or warm) and cold water, ideally from a running supply
- ☑ soap or other cleansers
- ☑ towels or another means of drying
- ☑ separate facilities for men and women except where they:
  - are in a separate room that can be locked from the inside
  - can only be used by one person at a time
  - are only used for washing the hands, forearms and face.

The provision of toilets must include separate facilities for men and women except where each toilet is in a separate room that can be locked from the inside.

What if my employees are working at temporary worksites?
'So far as is reasonably practicable' you need to provide flushing toilets and running water. Portable toilet facilities are available from hire companies, from those that require plumbing in, to self-contained units that come with their own generator and water supply. If this is not possible, consider alternatives (such as chemical toilets and water containers). Retro fitting of hand washing facilities to vehicles is also possible.

Use of public toilets and washing facilities should be a last resort and not used just because they are the cheaper option. This would not be acceptable where the provision of better facilities would be reasonably practicable.

Health and welfare

Food safety

The Food Safety (General Food Hygiene) Regulations apply to all workplaces, including building or construction sites, where food or drink is supplied, provided or sold for the benefit of employees and others working on site. They do not apply to sites where employees or other people only consume their own food and drink.

These regulations only set out basic hygiene principles. They focus on how to identify and control food safety risks at each stage of the process of preparing and selling food.

The regulations are *goal setting*, which means that, rather than prescribing defined actions that should be taken, they allow for food safety risks to be assessed and the appropriate controls to be put in place. Controls do not have to be complex; a simple example would be the use of refrigeration to prevent the growth of harmful bacteria on perishable food.

The person in control of supplying food for the consumption of others must:

- ✓ make sure food is supplied or sold in a hygienic way
- ✓ identify food safety hazards
- ✓ know which steps in the activities are critical for food safety
- ✓ ensure safety controls are in place, maintained and reviewed.

The location, design and construction of the premises must aim to avoid the contamination of food and harbouring of pests. It must be kept clean and in good repair in order to avoid food contamination.

Surfaces in contact with food must be easy to clean and, where necessary, disinfected. This will require the use of smooth, washable, non-toxic materials.

Adequate provision must be made for cleaning foodstuffs, and the cleaning and (where necessary) disinfection of utensils and equipment. All possible steps must be taken to avoid the risk of contamination of food or ingredients.

Food handlers must maintain a high degree of personal cleanliness, work in a way that is clean and hygienic and wear clean and, where appropriate, protective over-clothes. Adequate changing facilities must be provided where necessary.

Food handlers must protect food and ingredients against contamination that is likely to render them unfit for human consumption or create a health hazard.

Anyone whose work involves handling food should:

- ✓ observe good personal hygiene
- ✓ routinely wash their hands before handling foods
- ✓ never smoke in food handling areas
- ✓ report certain illness (like infected wounds, skin infections, diarrhoea or vomiting) to their manager or supervisor immediately, and stay off work if necessary.

Food handlers must receive adequate supervision, instruction and training in food hygiene. Each food business must decide what training or supervision their food handlers need by identifying the areas of work most likely to affect food hygiene.

B 08

Health and welfare

B 08

09
Personal protective equipment

What your employer should do for you	132
What you should do as a supervisor	133
Introduction	134
Duties of employers	134
Duties of employees	136
Risk management	136
Counterfeit PPE	137
Using PPE in practice	137
Differing standards	137
Selecting the correct PPE	138
Types of PPE	139
Caring for PPE	144

Personal protective equipment

What your employer should do for you

1. Put avoidance and engineering controls in place to avoid the need for personal protective equipment (PPE) where possible.
2. Assess the need for PPE, engaging the assistance and advice of reputable suppliers if necessary.
3. Ensure everyone on site has been issued with the PPE that they require to carry out the jobs allotted to them.
4. Provide training and instruction in the proper use and care of PPE.
5. Ensure everyone on site knows how and where to obtain any extra PPE that they might need.
6. Provide facilities for everyone on site to obtain replacement PPE for that which is lost or defective.
7. Maintain a PPE issue log.
8. Ensure that all PPE is obtained from a reputable supplier to avoid the inadvertent use of non-effective counterfeit items.

Personal protective equipment

What you should do as a supervisor

Checklist	Yes	No	N/A
1. Check that avoidance and engineering controls are in place and working to avoid the need for personal protective equipment (PPE) where possible.			
2. Monitor the need for PPE, engaging the assistance and advice of reputable suppliers where necessary.			
3. Ensure everyone on site has been issued with the PPE that they require to carry out the jobs allotted to them.			
4. Provide training and instruction in the proper use and care of PPE.			
5. Ensure everyone on site knows how and where to obtain any extra PPE that they might need.			
6. Provide facilities for everyone on site to obtain replacement PPE for that which is lost or defective.			
7. Maintain a PPE issue log.			
8. Check that all PPE is obtained from a reputable supplier to avoid the inadvertent use of non-effective counterfeit items.			

B 09

Introduction

Personal protective equipment (PPE) is any item of equipment or clothing that is used or worn by a person to protect them from an identified risk to their health or safety. A sub-group of PPE that is designed to protect the wearer against respiratory (breathing) hazards (such as the inhalation of dust and fumes) is known as respiratory protective equipment (RPE).

For simplicity, throughout this chapter, each reference to PPE will also include RPE unless indicated otherwise in the text.

Within the construction industry PPE is commonly thought of as equipment that is used to protect the head, ears, eyes, respiratory (breathing) system, skin, hands and feet.

However, it must be appreciated that PPE is also commonly used during construction activities to:

- ✓ prevent or arrest falls (for example, harnesses and lanyards)
- ✓ enable a person in distress to be rescued from a confined space (for example, a rescue harness)
- ✓ enable someone who has fallen into water to stay afloat and be rescued (for example, a lifejacket).

All PPE for use at work should be selected by a competent person who can ensure, in discussion with the supplier, that it meets the appropriate standards.

PPE should bear the CE mark to show conformity with European Standards.

Duties of employers

Health and Safety at Work Act 1974

The Act requires that employers, when providing anything in the interests of health or safety (for example, PPE), provide it free of charge.

Management of Health and Safety at Work Regulations

These regulations require that employers carry out suitable and sufficient risk assessments of the work they do and put in place measures to control the risks arising from the work. Part of this process will require that a safe system of work is developed, which in some circumstances will necessitate identifying and providing suitable PPE.

For further information refer to Chapter A04 Risk assessments and method statements.

Control of Substances Hazardous to Health Regulations

The regulations require that employers identify the hazardous properties of substances that are used at work or created by a work process. Control measures must be put in place to protect the health of anyone who would otherwise be affected by the hazardous properties of those substances. The use of PPE may be selected as a control measure although only after the implementation of other, more effective, control measures has been explored and found not to be reasonably practicable.

Personal protective equipment

There are many substances that are used in the construction industry that have been identified as being potentially harmful if they are inhaled or come into contact with exposed skin. The fact that many construction workers suffer from occupational asthma and/or dermatitis, with numbers increasing annually, shows that the risks are not being properly controlled. Part of this is the failure to provide and/or wear PPE where it is necessary.

The importance of the correct PPE being provided and used, when risks cannot be controlled by other means, cannot be over emphasised

 For further information refer to Chapter B12 Hazardous substances.

The following sets of regulations require that appropriate PPE is supplied and worn where an assessment shows that there is a risk of exposure to the respective hazards.

Construction (Head Protection) Regulations	see this chapter.
Control of Asbestos Regulations	see Chapter B10.
Control of Lead at Work Regulations	see Chapter B12.

Personal Protective Equipment at Work Regulations
These regulations require that employers:

- ☑ provide suitable PPE for their employees and make sure that it is used properly
- ☑ make sure that items of PPE are compatible when more than one item is worn at the same time (for example, if wearing ear defenders, it is possible that the headband would interfere with the correct fit of a safety helmet)
- ☑ make an assessment of the most suitable PPE to protect against the identified risks
- ☑ make sure that PPE is properly maintained where this is necessary
- ☑ replace PPE that is damaged or lost
- ☑ provide suitable accommodation where necessary for PPE that is not in use
- ☑ provide employees with adequate information, instruction and training on:
  - the risks that the PPE will avoid or limit
  - why the PPE has to be worn and how it should be used
  - how to maintain the PPE in efficient working order and good repair.

B 09

Duties of employees

In the context of PPE the Health and Safety at Work Act 1974, requires that employees:

- ☑ look after their own health and safety
- ☑ follow their employer's safe systems of work, including using anything provided for their health or safety as instructed.

The Personal Protective Equipment at Work Regulations, place duties on employees to:

- ☑ use any PPE provided in accordance with the instruction and training provided
- ☑ report to the employer the loss of or defect in any PPE provided.

Employees must follow their employer's safe systems of work

Risk management

After carrying out a risk assessment and establishing the hazards associated with a particular job, the employer must then implement measures to control the risks to health and safety. This may involve the identification and issue of appropriate PPE.

 PPE must only be selected as a means of controlling risks as a last resort.

All other methods of controlling the risks arising from the work activity must have been considered and found not to be reasonably practicable before the decision is taken to rely upon PPE for protection.

It must be remembered that for PPE to be fully effective, the user must have received adequate training and instruction in its use and it must:

- ☑ have been designed to protect the user against the type of hazard that will be present
- ☑ be available at all times that it is needed
- ☑ be adjusted properly where necessary
- ☑ fit the wearer properly and be compatible with other PPE worn at the same time
- ☑ be worn/used during the period(s) of risk
- ☑ be treated with care and returned to its storage after use, where this is necessary
- ☑ be inspected and maintained as necessary
- ☑ be replaced if it becomes defective.

Failure of an item of PPE, or using the incorrect type of PPE, could expose an employee to the possibility of serious injury, ill health or even death.

Fall arrest systems are a type of PPE

Counterfeit PPE

Choose only products which are CE marked in accordance with the PPE regulations.

 Always buy your PPE from a reputable supplier.

Using PPE in practice

Construction industry workers will have the need to wear or use PPE on many occasions. They will wear or use it for one of two reasons:

- ☑ because they have been told to (site rules), or
- ☑ because it makes sense.

The more often that it is done for the second reason, the better. In circumstances when wearing or using PPE is necessary it must become second nature for those workers who are at risk. However, at present this is still often not the case.

- ☑ Cases of occupational asthma and dermatitis show that PPE that protects the skin and respiratory system are not being used where they should. The problem is not being taken seriously either by employers, supervisors or employees.
- ☑ Deaths have occurred through falls, either because a harness and lanyard were not being worn or because they were worn but the free end of the lanyard was not clipped onto a suitable anchorage point.

There is the temptation to ignore the need to wear PPE and the protection it gives because 'the job will only take a minute' …

… and that 'minute' may be all the time that the job needs to kill or injure someone.

Differing standards

The requirements for wearing PPE vary greatly in the construction industry. For example, many companies have mandatory requirements for wearing gloves (even of different types to match the task) and light eye protection (safety specs) as a recognition of the need to minimise injuries to their workforces' hands and eyes, while working in an ever changing work environment.

Personal protective equipment

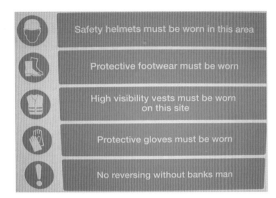

Different sites will have different requirements

Selecting the correct PPE

 How do I choose the correct PPE (size, fit and type)?

In asking this question, the employer must ensure that certain factors are taken into consideration.

The task

 Which risks to health and safety, that can only be controlled by the wearing or use of PPE, arise from this job?

The PPE selected must offer the level of protection required for the hazard(s) that have been identified. For example, will the job involve:

- ☑ the use of substances that have the potential to create hazardous dust or fumes?
- ☑ the use of substances that could irritate or burn the skin?
- ☑ creating airborne dust through cutting, grinding and so on?
- ☑ creating fumes from hot-work processes?
- ☑ any process that could result in eye injuries?
- ☑ working at height in circumstances where the wearing of a harness and lanyard is the only practical fall protection measure?

 As well as assessing PPE for the task you must also assess any additional PPE requirements for the surrounding site conditions or from others (such as noise, dust, fumes, falling objects and so on).

PPE must be appropriate to site conditions

Personal protective equipment

Manufacturers and suppliers have a statutory duty to provide information regarding the performance characteristics of the PPE products that they manufacture/sell. If necessary, they should be consulted.

 If you cannot prevent, you must protect.

The fit/individual

For some basic PPE (such as a pair of general safety spectacles), the exact fit may not be an issue. However, in selecting some PPE, achieving a satisfactory fit is essential. For example:

- ☑ when a respirator is to be used, the effectiveness of the device is dependent upon the face fit, which in turn will require that the head-harness straps be adjusted to suit the wearer and facial features *(refer to Chapter B10 Dust and fumes (Respiratory risks))*
- ☑ adjusting the head-harness of a safety helmet to suit the wearer will ensure that it is comfortable and secure
- ☑ gloves should be a close fit to protect from chemical ingress, entanglement and to maximise dexterity (buying gloves in one size only should be avoided)
- ☑ adjusting the fit of a safety harness, which could at worst have to take the shock loading of an arrested fall, is essential if the harness is to be fully effective.

Where a satisfactory fit is not, or cannot, be achieved, the wearer is likely to suffer discomfort and is much more likely to stop using the PPE and also lose concentration regarding the job in hand.

Co-operation between the employer, supervisor and employee must be established in the selection of comfortable and acceptable PPE.

Consider the potential wearers' physical needs. If they wear glasses and have to wear a mask, goggles or safety glasses, they may require a special type. If they have a beard and have to wear respiratory equipment, can they obtain an airtight seal? These are crucial considerations.

 Protection is available for the hazard that you are about to work with. Find it and use it.

The reasons

Supervisors must ensure that employees understand 'why', 'what', 'who', 'how' and 'when' there is a need for PPE.

- ☑ **Why** is the PPE needed?
- ☑ **What** will be the implications of not wearing it?
- ☑ **Who** is going to provide the PPE along with all necessary information, instruction, training and supervision?
- ☑ **How** is it fitted, worn or adjusted?
- ☑ **When** must it be worn?

Types of PPE

There is a vast range of PPE available to not only provide protection from numerous hazards but personal comfort as well. The range covers, for example:

- ☑ overalls
- ☑ coveralls
- ☑ work wear tops and trousers
- ☑ knee pads

Personal protective equipment

- ✓ lifejackets
- ✓ high and low temperature clothing
- ✓ hard hat liners
- ✓ wet weather protection.

Eye protection

Eye protection is required by law when there is any possibility of eye injury due to, for example, grinding, welding, cutting, hammering or working with hazardous fluids. The risks include:

- ✓ impact by solids
- ✓ ingress of liquid, dust or gas
- ✓ splashes of hot metal
- ✓ exposure to harmful forms of light.

Depending upon the nature of the hazard some of the features of eye protection will be:

- ✓ impact or heat-resistant lenses or screen
- ✓ ventilated facepiece to stop the lenses misting
- ✓ an effective gas or liquid-proof seal around the face
- ✓ chemical-resistant lenses or screen
- ✓ lenses or screen that filter out harmful light (such as welding flash or laser light).

Two common types of eye protection worn during construction activities are safety spectacles (including light eye protection and safety goggles) and face shields.

There will always be a type of eye protection available that will meet your particular needs. If there is any doubt about the best type, BS EN 166 or the supplier should be consulted.

It should be noted that light eye protection (safety spectacles) are not impact rated and should therefore not be used for tasks that require impact protection (such as operating grinders, cut-off saws, nail guns and so on).

In case of eye injury:

- ✓ no medication should be applied to the affected eye
- ✓ cover the eye with clean, dry materials and seek immediate medical attention.

Head protection

The Construction (Head Protection) Regulations have been withdrawn. The requirement to wear suitable head protection on all building and construction sites, unless there is no risk of head injury, either from falling objects or from banging the head, is now covered by the PPE Regulations.

Ensure that head protection is always worn by everyone on site, except when in designated safe areas (such as the site office or canteen).

There are many makes of safety helmet available that are constructed to British Standards, and it is the duty of the employer to provide safety helmets that are suitable for the job and the wearer.

Employees, for their part, must follow their employers' instructions and wear the safety helmets at all times when instructed to do so, and report any damage or loss to their employer.

Personal protective equipment

Safety helmets are designed to offer a pre-determined level of impact resistance if they are correctly worn.

 The practice of wearing safety helmets back-to-front must not be tolerated.

It has been established that the solvents in some paint, adhesives and indelible markers can reduce the strength of the plastic from which helmets are manufactured. Employees must be discouraged from marking or otherwise decorating their safety helmets other than applying official stickers (such as 'First aider') or confirmation of site induction.

A safety helmet that has fallen from height onto a hard surface may have suffered damage that will affect its strength even though no cracks are visible. In most circumstances, a replacement helmet should be obtained.

Where the work involves leaning over exposed edges, or similar, chinstraps must be fitted and worn. Many safety helmets have in-built features that enable compatible ear defenders or a face shield to be securely attached.

Sikhs

Under Regulation 11 of the Employment Act, construction workers who are practising members of the Sikh faith are exempt from wearing a safety helmet whilst wearing a turban. No other workers are covered by this exemption. Sikhs who are not wearing turbans are not exempt from the regulations and, therefore, are required to wear the same head protection as other operatives.

Hearing protection

 Refer to Chapter B11 Noise and vibration.

Respiratory protective equipment

 Refer to Chapter B10 Dust and fumes (Respiratory risks).

Hand protection

There are many types of glove available to protect the wearer against the different types of hazard that can be present in the workplace (such as cuts, abrasions, chemicals, heat, cold and other hazards).

Employers must provide the correct type of glove for the job as the wrong type will often offer no or little protection or even make the matter worse.

Key points

 Assess:
- task
- application (job task risks of wearing or not wearing gloves)
- wearer needs
- suitability of gloves for the job, matching type and size.

 Once a decision has been made:
- record assessment
- brief the work team.

Personal protective equipment

Hand protection must be suitable for the job

Employers and supervisors need to understand that if gloves are not right for the job and create difficulty in use they are not likely to be worn. There may be situations in which gloves that provide the required level of hand protection do not allow the level of 'feel' to do the job and there may be the temptation to work without them or, if worn, they could even pose a hazard (such as entanglement) when using machinery.

Gloves will wear with use and it is important that a supply of replacement gloves are readily available or, inevitably, someone will end up working without wearing them. Even a small hole in a glove that is worn to protect the user against a hazardous liquid will make it ineffective.

In this case, it will be necessary to contact a PPE supplier or consult a PPE catalogue to investigate the types of glove that are available that will satisfy both needs. Modern materials are enabling the production of gloves that are both robust and permit a high level of feel or dexterity.

Foot protection

Accidents arising from the manual handling of articles and substances are common causes of injury to the feet. Upward pointing sharp objects (such as nails sticking through pieces of wood) also have the potential to cause foot injuries.

It is common for sites to impose a 'no boots – no job' policy with regard to wearing safety footwear.

Foot protection comes in many types and styles, from safety trainers, safety shoes and boots and safety wellington boots to rigger boots.

- ☑ **Safety trainers** offer good grip on sloping or slippery surfaces and offer more comfort. They are more suitable to trades (such as floor layers) who repeatedly kneel and bend their feet.
- ☑ **Safety wellington boots or waders** are essential in preventing burns when operatives have to stand in wet concrete. The cement content, when mixed with water, becomes highly corrosive and will cause severe burns to body tissue.
- ☑ **Rigger boots**, as commonly worn by construction workers, provide the required level of protection with steel toecaps and a steel plate moulded into the mid-sole, protecting the wearer against dropped objects and penetration through the sole by sharp objects.

Safety footwear can also be oil and slip resistant. Suppliers should be consulted if there is any doubt about the type of footwear that is available for specialist work activities.

The ankle support provided by some styles of safety boot is important in the prevention of injuries resulting from walking on uneven surfaces and some industries insist on this type (for example, the rail industry).

Personal protective equipment

Hi-visibility clothing

Whether working on site or just visiting, site rules on most sites require all personnel to wear a high-visibility (hi-vis) jacket, waistcoat (vest) or overalls at all times.

There are several classes of hi-vis clothing used and full specifications can be found in BS EN 471. In addition there are a number of local and specific regulations, for example:

- ☑ slinger signaller/banksman
- ☑ first aiders
- ☑ working on or near highways, roads and streetworks
- ☑ the rail industry.

Rail work requires specific hi-vis clothing

Roads and streetworks

Class 1	Defines the lowest visibility level. Example: high-visibility trousers with two 5 cm reflective bands around each leg. These become Class 3 when worn with a Class 3 jacket.	The colour of the background material should normally be fluorescent yellow from Table 2 of BS EN 471, and the reflective material should comply with Table 5 of BS EN 471.
Class 2	Defines an intermediary visibility level. Example: vests. Two 5 cm bands of reflective around body or on one 5 cm band around body and braces to both shoulders.	Colour scheme as Class 1.
Class 3	Defines the highest visibility level. Example: jacket with long sleeves, jacket and trouser suit. Two 5 cm bands of reflective tape around the body, arms and braces over both shoulders.	Colour scheme as Class 1. Must be worn on dual carriageway roads with a speed limit of 50 mph or above and must comply with the colour scheme to BS EN 471.

Supervisors and wearers must ensure the garment is in good condition and is properly fastened at the front and not modified (for example, cutting the bottom band off a hi-vis vest must not be allowed).

Personal protective equipment

Harnesses

For information on fall protection PPE, for the implications of apparently minor damage to fall arrest PPE, including working over or near water refer to Chapter D20 Working at height.

Caring for PPE

The users of PPE are generally responsible for looking after it on a day-to-day basis. For simple items of PPE (such as safety spectacles), this might only involve keeping the lenses clean and free of scratches. For more complex PPE (for example, a safety harness) there will be a need for a:

- ☑ visual inspection each time before it is used
- ☑ schedule of periodic, detailed and recorded examinations by someone who has been trained to do so.

The information supplied by the manufacturers of PPE should outline their recommendations on the need for inspections, maintenance and the general care of their products, including where necessary, there is a need for specialist knowledge.

For further information refer to GE 700 *Construction site safety,* **Chapter B06 Personal protective equipment.**

Storage of work clothes and/or personal clothes may need to be provided

10

Dust and fumes (Respiratory risks)

What your employer should do for you	146
What you should do as a supervisor	147
Introduction	148
Respiratory (breathing) diseases	148
Respiratory protective equipment	150
Asbestos	153
Pigeons	159
Silica	159
Silicosis	160

Dust and fumes (Respiratory risks)

What your employer should do for you

1. Ensure that dust, fumes and vapours are either eliminated or minimised.
2. Provide systems and equipment to ensure that airborne dusts are at the lowest level to minimise the risk of asthma and other lung diseases.
3. Carry out, or arrange to have carried out, a refurbishment or demolition survey for remedial and refurbishment work, and for demolition due, prior to work starting.
4. Ensure that information about asbestos is made available to the workforce and others who may be affected.
5. Provide information on exhaust systems/wet cutting methods or RPE to protect dust exposure to the workforce.
6. Issue workers with, and train them in the use of, the correct RPE that they require to carry out their jobs safely.

Dust and fumes (Respiratory risks)

What you should do as a supervisor

Checklist	Yes	No	N/A
1. Check that dust, fumes and vapours are either eliminated or minimised by the proper use of equipment provided.			
2. Check that systems and equipment are working to ensure that airborne dusts are at the lowest level to minimise the risk of asthma and other lung diseases.			
3. Ensure that the work team are fully briefed on any asbestos refurbishment or demolition survey for remedial and refurbishment work prior to work starting.			
4. Check that training and information about asbestos is made available to the work team and others who may be affected and that they understand the importance of stopping if they are not sure.			
5. Monitor effectiveness of exhaust systems/wet cutting methods and RPE to protect dust exposure to the work team.			
6. Ensure that your work team are trained in the use of and have the correct RPE, that they use it and report defects.			
7. Monitor that RPE is worn and is used correctly.			

B10

Dust and fumes (Respiratory risks)

Introduction

- ☑ It is estimated that there were at least 148 new cases of occupational asthma in 2011.
- ☑ 4,000 people die each year from asbestos-related disease.
- ☑ 800 people (and more each year) are dying from silica-related cancers.
- ☑ Many more suffer life-changing illnesses.
- ☑ 1.7 million days are lost to work-related ill health each year.

 At any one time there are far more people off work through occupational ill health than there are because of a work-related accident.

Respiratory (breathing) diseases

Exposure to everyday hazards, including flux, welding and cutting fumes, dusts, asbestos and many more everyday workplace materials and processes can cause ill health to the workforce.

The breathing in of dusts and fumes, known as respiratory sensitisers, is likely to cause an allergic reaction. As an industry we are exposed to many materials and products and are therefore exposed to particular dust hazards (such as asbestos, cement, stone, silica, lead, fillers, MDF, plastics, epoxys, solvents and so on).

The effects of exposure may be immediate or it may take years for symptoms of ill health to become apparent. The inhalation of hazardous airborne contaminants can cause wheezing, coughing, breathlessness, bronchitis and other respiratory diseases, including various types of cancer.

Stomach disorders may be brought on by the ingestion of the dust of some substances, due to eating food with dirty and contaminated hands, or simply eating in a dusty environment.

The main respiratory hazards that may be encountered on site are shown below.

- ☑ **Mists:** tiny liquid droplets formed by the atomisation of a liquid (for example, when spraying or using an aerosol). Mists may be a combination of several hazardous substances.
- ☑ **Metal fumes** occur when metal is heated to high temperatures, (such as during welding and gas cutting). Fumes contain minute metal particles that may remain in the air, and be inhaled, for some time.
- ☑ **Gases:** airborne at room temperature. These normally mix with the air that we breathe (for example, propane, butane, acetylene, carbon monoxide, hydrogen sulphide), and can spread very quickly.
- ☑ **Vapours:** the gaseous state of substances that are liquids or solids at room temperature. They usually form when substances evaporate (for example, the vapour from a tin of glue or solvent left open).
- ☑ **Dusts** produced when solid materials are broken down into finer particles. The longer that the dust stays in the air the easier it is to breathe in.

Dust is by far the most common hazard on site.

What is dust?

Dust can simply be described as particles in air. What it consists of is purely down to what material is being cut, sanded or drilled, as the physical action of the tools breaks off small parts of the material into the atmosphere.

Dust and fumes (Respiratory risks)

The harm that it can do is dependent on what the material is.

 Don't fall into the misconception that all dust is the same.

Common materials that can do particular harm are shown below.

- ☑ **Wood – hardwood, softwood and plywood** dust is a serious problem. It can cause allergic reactions or cause cancers in the nose and lungs. Softwood dust is known to cause sensitisation – an allergic reaction – whereas hardwood is a known carcinogen (causes cancer).

- ☑ **MDF** – made from separated softwood fibres, hardwood fibres and glues. When cut it has the greater potential to form a particularly fine dust. This increases the likelihood that dangerous amounts can be breathed in.

- ☑ **Stone, brick, block, sand and concrete** – all contain a compound called silica. Very fine silica dust is raised when concrete, bricks, blocks, tiles and stone are sanded, cut or drilled. The effects of breathing in silica dust are not quick, and may take 10 to 30 years to become apparent. Silicosis is the lung disease that occurs and can be fatal. There is no cure.

Prevention

Here are three simple steps that should be taken to protect workers and the people around them.

- ☑ **Avoid creating dust**. Choosing the right equipment or method of work can really protect workers' lungs and potentially eliminate the risk altogether. Pre-order sized materials rather than cutting them on site and using a block splitter rather than a disc cutter creates no dust and is quicker.

- ☑ **Stop the dust getting into the air**. If creating dust can't be eliminated then stopping and minimising dust being released should be the priority every time. This can be done two ways.
  - **Dampening down or wet cutting.** This is the cheapest and most effective way. Water helps form a slurry preventing the majority of dust becoming airborne. Not only does it reduce what is breathed in, but it also has the benefit of less cleaning up afterwards.
  - **Capturing the dust.** Materials (such as wood) do not suit the use of water to suppress dusts so consider alternatives (such as dust extraction). When purchasing or hiring tools, ensure they have the facility to extract the dust, as this is the only effective way to capture the dust being released. Some tools are fitted with dust bags, but these have only limited efficiency and clog up easily. Always try to use a means that will remove the dust by suction and regularly clean filters.

- ☑ **Wear protection**. Even the best control measures won't prevent all dust being released, so respiratory protection equipment (RPE) or dust masks must always be worn as well. Dust masks with no filter rating (and usually only having one strap) offer little or no protection against microscopic hazardous dust.

Further measures will include:

- ☑ the provision of information, instruction and training to employees
- ☑ the strict observance of all recommendations and procedures advised by the manufacturer
- ☑ the effective supervision of employees, and the monitoring of work methods and practices
- ☑ the provision of protective clothing and equipment before any work starts

Dust and fumes (Respiratory risks)

- [✓] the correct disposal of waste materials and containers as recommended by the manufacturers
- [✓] cleaning (for example, by extracting dust using a vacuum cleaner, rather than stirring it up by sweeping)
- [✓] personal hygiene, including the cleansing of hands before consuming food, the use of barrier creams, the removal and storage of contaminated clothing during meals, and the correct laundering or disposal of contaminated clothing
- [✓] to avoid, where practical, carrying out potentially hazardous work activities in confined areas to lessen the chances of dust concentrations or fumes building up.

Using a water-fed cut-off saw whilst wearing respiratory protection

Exposure limits

The accurate measurement of exposure to airborne substances is a specialist activity.

Material safety data sheets can be used to carry out some risk assessments on how and where materials are to be used. If in any doubt seek expert advice.

Manufacturers offer excellent advice and guidance on selection and use of respiratory protection equipment.

Exposure to any general nuisance dust should be limited by reducing dust levels to the minimum reasonably practicable and should not exceed 10 mg of dust per m^3 of air. Generally speaking, if dust is visible in the air, it is highly possible that the 10 mg limit is being approached (or possibly exceeded), and the application of the Control of Substances Hazardous to Health (COSHH) Regulations should be considered.

Occupational asthma, when caused by exposure to any substance known to be a respiratory sensitiser is a notifiable disease under the Reporting of Injuries, Diseases and Dangerous Occurrences Regulations (Revised) (RIDDOR). If in any doubt, contact your local HSE office or the HSE Employment Medical Advisory Service (EMAS).

Respiratory protective equipment

Respiratory protective equipment (RPE) protects the user against the risk of respiratory illness, often providing protection against unseen hazards (such as fine dust, invisible fibres, gases and vapours). It is essential that the correct type of RPE, or even the correct type of filter cartridge for a respirator, is selected.

RPE works by providing the wearer with air that is safe to breathe by:

- [✓] filtering out the contamination from the air in the workplace (for example, simple, disposable filtering half-masks), or

Dust and fumes (Respiratory risks)

- supplying a stream of fresh air from an external source (for example, self-contained breathing apparatus).

To ensure the effectiveness of any RPE, which depends upon the seal between the facepiece and the user's skin, face-fit testing must be carried out by a trained person. The effectiveness of the seal will depend upon a number of factors, including the shape of the face and facial hair. Face-fit testing must allow for any strenuous activities or exertions that occur whilst the RPE is being worn.

Types
Some examples of RPE are:

- disposable filtering half-masks
- half-mask respirators with detachable filter cartridge
- ventilated visor or helmet respirators.

For other hazardous situations, breathing apparatus fed with breathable air from either compressed air bottles or a fresh air hose, are available. Training is required in the use of this equipment.

It should be noted that nuisance dust masks (simple gauze filters or cup-shaped filters often held in place by a single strap) are not classed as PPE or RPE. They do not meet any current standards of protection or legislative requirements.

Selection
The choice of RPE will depend upon the nature of hazard from which protection is required. In many cases RPE will only protect against one type of hazard (for example, dust or fumes). A wide range of RPE is available and further guidance on the selection of suitable RPE is shown below and available in:

- the HSE publication *The selection, use and maintenance of respiratory protective equipment* (HSG53)
- the guidance document *Respiratory protective devices. Recommendations for selection, use, care and maintenance* (BS EN 529:2005).

RPE must only be selected by a person who has a thorough knowledge of the hazard and the types of RPE available. It must be the correct RPE for the job.

Selecting the wrong type of RPE could have serious, even fatal, consequences. Selection must be carried out by a competent person. Some of the factors that will determine the appropriate type of RPE are the:

- hazardous nature of the substance
- airborne concentration of the substance
- period of exposure
- wearer's required field of vision
- provision for communication
- need to move in cramped or difficult working places
- prevailing weather conditions
- suitability of the protective equipment for the individual
- need for an external source of breathable air.

Dust and fumes (Respiratory risks)

Notes:

1. *When selecting suitable respiratory protective equipment (RPE) it may be necessary to seek expert advice from manufacturers/suppliers.*
2. *Training in the types of, and in the use of, respiratory equipment must be given.*
3. *A face-fit test is required for any tight fitting facepiece – this will include all disposable masks, half masks and most powered masks.*
4. *All masks, other than disposables, require regular examination by a competent person and records to be kept.*
5. *Guidance and recommendations are provided in the HSE publication Respiratory protective equipment at work – A practical guide.*

Some types of respiratory protective equipment (RPE) that are used in industry are:

- ☑ disposable face mask respirators
- ☑ half-mask dust respirators
- ☑ powered respirators
- ☑ ventilated visor and ventilated helmet respirators
- ☑ compressed air line breathing apparatus
- ☑ self-contained breathing apparatus
- ☑ full face masks.

Each type of RPE is given an assigned protection factor (APF), which gives the user some idea of the level of protection that the device will provide. For example, a disposable mask marked:

FFP1 or P1	offers a protection factor of 4
FFP2 or P2	offers a protection factor of 10
FFP3 or P3	offers a protection factor of 20.

The APF assumes that the user has passed a face-fit test, is wearing the mask correctly, is clean shaven and there is no other interference in fit (such as facial piercing).

The APF is a number that can mean two things. For example, a particle mask with an APF of 10 means either that:

- ☑ for every 10 units of contaminant outside the mask, only one unit will get inside the mask, so the mask filters out nine out of 10 of the units
- ☑ or, if there is a time exposure limit, then a mask extends the time by the same factor – so if someone can be exposed to a concentration of a contaminant for 10 minutes without needing to wear a mask, then a mask with an APF of 10 would allow the person to be exposed for 100 minutes (10 x 10).

This is a very basic introduction into the theory behind the choice of RPE.

Manufacturers' product literature also give very good RPE selection and use guidance.

Common RPE failings

Recent research undertaken by the HSE targeted workplace RPE inspections and found:

- ☑ a perception that using RPE was simple/obvious
- ☑ managers and supervisors had knowledge gaps on critical aspects of RPE
- ☑ there was limited use of external information sources (for example, manufacturers' literature and websites)
- ☑ there was an assumption that workers knew how to put on and use RPE

Dust and fumes (Respiratory risks)

- ✓ RPE misuse is common – but it is generally blamed on the workers
- ✓ limited management input/involvement
- ✓ little information was provided to the worker
- ✓ poor standards of training, hazard awareness and RPE use
- ✓ little supervision and enforcement
- ✓ deficiencies in provisions for RPE storage and maintenance regimes
- ✓ RPE pre-use checks unlikely.

In addition, other failings include:

- ✓ inadequate face-fit testing
- ✓ wearers having more than one day's stubble growth, which substantially reduces the RPE's seal around the face and thus the protection factor
- ✓ mask straps not being adjusted, tightened or straps in the wrong position on the head.

For further information refer to GE 700 *Construction site safety*, Chapter B10 Dust and fumes.

Asbestos

The HSE has revised the ACoP which supports The Control of Asbestos Regulations. For further information refer to www.hse.gov.uk/asbestos

Asbestos is a naturally occurring fibrous material and was widely used in the UK until it was banned in late 1999. It is used as an insulator (to keep in heat and keep out cold), has good fire protection properties and protects against corrosion.

Because asbestos is often mixed with another material, it is hard to know if you're working with it or not. But, if you work in a building built before the year 2000, it is likely that some parts of the building will contain asbestos.

Asbestos may be found:

- ✓ as a pure substance (for example, as fibrous lagging around pipework)
- ✓ as a constituent of a mixture of materials (for example, mixed with other substances in floor or ceiling tiles)
- ✓ in any structure built before 1999.

If an unknown substance is discovered that could be or could contain asbestos, work must stop immediately and measures taken to keep the area clear of people.

Always assume that an unknown substance is asbestos until it is confirmed that it is, or it is proved to be something else. The removal of asbestos, or even taking samples of suspected asbestos for analysis are specialist activities, which must be left to specialist contractors.

Asbestos sample analysis is not expensive and results can be obtained relatively quickly.

Dust and fumes (Respiratory risks)

The identification of any substance thought to be or contain asbestos can only be established by laboratory analysis.

There are three main types of asbestos:

- ☑ blue asbestos – otherwise known as crocidolite
- ☑ brown asbestos – otherwise known as amosite
- ☑ white asbestos – otherwise known as chrysotile.

The colour of the asbestos that is discovered will often be misleading. The colour may have changed through ageing or by the application of heat, paint or as a result of being encapsulated in other materials. Original building plans or specifications may confirm the presence of asbestos and give details about its type.

Asbestos waste must carry a warning label. The label can also be used to identify and manage known in situ asbestos.

The cancers and respiratory diseases that result from exposure to asbestos are notifiable under the Reporting of Injuries, Diseases and Dangerous Occurrences Regulations (Revised) (RIDDOR). If in any doubt, talk to your employer.

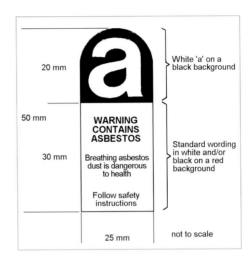

The Control of Asbestos Regulations

Asbestos was widely used as a building material in the UK until it was banned in 1999. Disturbance of the fabric of any building built before this date has the potential to expose those doing the work, and possibly other people, to asbestos. If you are a contractor working on someone else's (non-domestic) premises where asbestos is likely to be present, you must be given access to the asbestos register, which should outline the location and condition of asbestos within the building. There is no legal requirement for privately owned domestic properties to have an asbestos register, however there should be one for Local Authority and housing association housing stock.

Dust and fumes (Respiratory risks)

 If you are organising or supervising the work and you have not been provided with the asbestos register you must arrange for an asbestos survey to be carried out by a competent person or organisation.

If you are not given access to the asbestos register, or there isn't one, and you suspect that the building is of an age where asbestos is likely to be present, you must not start any work until a survey has been carried out.

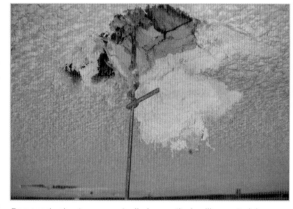

Damaged asbestos spray-applied acoustical ceiling material

 Under guidance outlined in the HSE publication *Asbestos: The survey guide* (HSG264), there are two types of survey:

- ☑ management survey
- ☑ refurbishment or demolition survey.

Where work is to be carried out that will disturb the fabric of the building, a refurbishment or demolition survey must be carried out. By necessity this must be an intrusive (invasive) survey, which attempts to locate asbestos in likely hidden locations (such as voids in partition walls, roof-spaces, underground ducts and so on).

When selecting an asbestos surveyor your employer must take reasonable steps to ensure their competency. If you appoint a surveying company they should be accredited to ISO 17020. The person who actually carries out the survey must:

- ☑ have sufficient qualifications, knowledge, experience and ability to carry out a survey and to recognise their limitations
- ☑ have sufficient knowledge of the specific tasks to be carried out and the risks involved
- ☑ be independent, impartial and have integrity
- ☑ carry out the survey in accordance with HSG264.

B 10

Dust and fumes (Respiratory risks)

Asbestos soffit

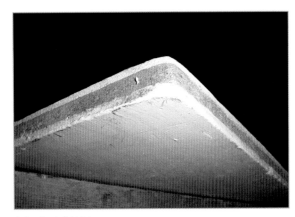

Asbestos loft hatch

Your employer must be specific when arranging a survey to ensure that there will be no limitations (caveats) to the surveyors' report that would leave you with uncertainty over the presence of asbestos in the area where work will be carried out.

The regulations direct the way in which asbestos must be handled, removed and disposed of. In summary the legislation:

- ☑ prohibits most work with asbestos, except by holders of licences issued by the HSE and from 6 April 2012, some non-licensed work needs to be **notified** to the relevant enforcing authority

- ☑ allows some work to be carried out by unlicensed contractors, providing certain time limitations are adhered to and/or if the work only involves certain materials that contain asbestos (such as asbestos-cement sheet), providing the requirements of the Control of Asbestos Regulations are complied with and where notifation has to be made to the HSE certain records kept

- ☑ requires that anyone in charge of premises effectively manages any asbestos that is situated in those premises

- ☑ prohibits the import, supply and use of asbestos in all new work and refurbishment/extension work, and so on

- ☑ stipulates the precautions necessary when asbestos is handled (such as the notification to HSE of any asbestos identified and its intended removal or maintenance)

- ☑ controls the way in which waste asbestos is managed.

The requirements of The Control of Asbestos Regulations apply to both licensed and non-licensed asbestos work.

Under Regulation 4, there is a duty on anyone who has control of non-domestic premises, either through ownership, terms of occupancy or a maintenance or repair contract, to manage any asbestos that is in those premises.

Dust and fumes (Respiratory risks)

 If workers disturb an unknown substance that could be asbestos, or contain it, they must stop work immediately and warn others to keep out of the area.

An employer whose employees carry out work with asbestos is required to:

- ✓ make a suitable and sufficient assessment about whether asbestos is, or is liable to be, present in the premises where work is being carried out
- ✓ identify the type of asbestos or assume the material contains asbestos
- ✓ assess the nature and degree of exposure, and the steps to be taken to reduce it
- ✓ prepare a suitable written plan of work
- ✓ notify the enforcing authority in most circumstances
- ✓ provide adequate information, instruction and training, for employees and others
- ✓ prevent or reduce asbestos exposure of employees to the lowest level reasonably practicable, by means other than the use of respiratory protective equipment (RPE)
- ✓ ensure the proper use of RPE
- ✓ maintain respiratory equipment in a clean, efficient state, good order and repair and regularly examine and test exhaust ventilation equipment
- ✓ provide adequate and suitable protective clothing and ensure that it is cleaned or disposed of appropriately
- ✓ prevent the spread of asbestos from the workplace

- ✓ ensure premises and plant involved in work with asbestos are kept clean
- ✓ designate, mark and limit entry into areas where exposure to asbestos exceeds, or is liable to exceed, the specified action level or control limit
- ✓ monitor the air where employees are exposed to asbestos and keep suitable records for a specific period
- ✓ ensure that any air monitoring carried out meets the required criteria
- ✓ ensure employees liable to be significantly exposed to asbestos receive regular medical surveillance by a medical employment adviser or appointed doctor
- ✓ provide washing and changing facilities that are adequate and suitable for employees exposed to asbestos, and storage for protective clothing and for personal clothing not worn during working hours
- ✓ regulate raw asbestos, asbestos waste storage and disposal, ensuring adequate packaging, sealing and marking in accordance with the regulations.

 Training is mandatory for anyone liable to be exposed to asbestos fibres at work. This includes maintenance workers and others who may come into contact with, or disturb, material (such as satellite dish installers), as well as those involved in asbestos removal works.

Dust and fumes (Respiratory risks)

Medical surveillance

Where employees' exposure to asbestos is likely to exceed the asbestos-in-air action level as defined in the Control of Asbestos Regulations, the employer has a duty to arrange medical surveillance for the affected employees. *(Refer to Chapter B08 Health and welfare for further information on medical surveillance.)*

By April 2015, all workers/self-employed persons doing notifiable non-licensed work with asbestos must be under health surveillance by a doctor. Workers who are already under health surveillance for licensed work need not have another medical examination for non-licensed work. Medicals for notifiable non-licensed work are not acceptable for those doing licensed work.

The duty to manage asbestos in non-domestic premises (Regulation 4)

This is perhaps the most important regulation for the long-term protection of the health of employees and others. It states that any person who owns, occupies, manages or has responsibility for premises (or a part of it) that may contain asbestos has either:

- ☑ a legal duty to manage the risk from asbestos or asbestos containing materials, or
- ☑ a duty to co-operate with whoever has the duty to manage the risk.

> The HSE publication *A short guide to managing asbestos in premises* (INDG223), explains this duty to manage in the following general terms.

If you have any maintenance and/or repair responsibilities for non-domestic premises, either through a contract or a temporary agreement, or because you own or occupy the premises, then you have a duty. This duty will require you to manage the risk from asbestos by:

- ☑ finding out if there is any asbestos or suspected asbestos containing materials (ACMs) in the premises, how much of it there is, where it is, and what condition it is in
- ☑ always presuming that it is asbestos, or an ACM, unless you know or you have strong evidence proving otherwise
- ☑ making and keeping up to date a record (an asbestos register) of the location and condition of all the asbestos, ACMs or presumed ACMs in your premises
- ☑ assessing the risk that the materials pose to employees and others
- ☑ preventing any work on the premises that may disturb asbestos or ACMs until control measures to manage the risk have been put in place
- ☑ preparing a plan that sets out in detail how you are going to manage the risk from the material
- ☑ taking all of the steps that you need to take to put your plan into action
- ☑ reviewing and monitoring your plan and the arrangements that you made to put it in place
- ☑ providing information on the location and condition of the material to anyone who is liable to work on or to disturb it
- ☑ ensuring that anyone who has information on the whereabouts of asbestos in your premises is required to make this available to you as a duty holder (those who are not duty holders but control access to the premises have to co-operate with you in managing the asbestos)

Dust and fumes (Respiratory risks)

☑ ensuring that any area from which asbestos has been removed has been thoroughly cleaned and certified as such before reoccupation.

The cancers and respiratory diseases that result from exposure to asbestos are notifiable under the Reporting of Injuries, Diseases and Dangerous Occurrences Regulations (Revised). If in any doubt, contact your local HSE office or the HSE Employment Medical Advisory Service (EMAS).

> For further information refer to GE 700 *Construction site safety*, Chapter B09 Asbestos.

Pigeons

Dry pigeon droppings or water droplets containing contaminated bird droppings, if disturbed, can become a hazardous airborne dust that can cause a severe respiratory illness. If work is to be carried out in an area where pigeons have been nesting or congregating, a risk assessment must be undertaken to identify the necessary control measures.

Activities such as cleaning windowsills will not result in high exposures to infected material and are not high risk. For larger quantities of dry droppings, the use of high pressure water should be avoided to minimise creating droplets of water, but wetting down the work area (using low pressure methods) will help to prevent inhalation of infected dust and reduce the risk of infection. It will also prevent the spread of dust outside the work area. Containing the work area with plastic sheeting should also be considered.

If required, as identified by the risk assessment, RPE may be required. For example when larger quantities of droppings are involved, a P3 or FFP3 mask should be used. Overalls should be worn when carrying out this work, and replaced when they are soiled. Good washing facilities should be provided and hygiene measures followed.

Silica

Silica occurs as a natural component of many materials used in construction activities. Crystalline silica is present in substantial quantities in sand, sandstone and granite, and often forms a significant proportion of clay, shale and slate. Products (such as concrete and mortar) also contain crystalline silica.

> The health hazards of silica come from breathing in the dust. Activities that can expose workers or members of the public to the dust include working with stone, grit blasting, scabbling, cutting or drilling, demolition and tunnelling.

The use of power tools leads to high exposures if exhaust systems/wet cutting processes are not used and maintained. For some activities, exposure will depend upon how confined the working space is, and the presence or absence of ventilation. (For example, tunnelling through dry, silica-bearing rock will always lead to high exposures for workers at or near the cutting face unless precautions are taken.)

Breathing fine dust of crystalline silica can lead to the development of silicosis. This involves scarring of the lung tissue and can lead to breathing difficulties. Exposure to very high concentrations over a relatively short period of time can cause acute silicosis, resulting in rapidly progressive breathlessness and death within a few months of onset.

Dust and fumes (Respiratory risks)

B 10

Materials	Sand, sandstone, granite, clay, shale and slate.
Common products	Concrete, mortar, bricks, blocks, ceramics, kerbs and paving slabs.
Common operations	Sweeping up, demolition and strip out. Drilling or breaking. Cutting (disc cutter/chasing machine/floor saw). Grinding, polishing, rubbing down, sanding and blasting.
Control measures	Manufacture off site/isolate process. Substitute process (for example, use hand block splitter not disc cutter). Extraction/vacuum collection/forced extraction. Local exhaust ventilation. Water suppression (typically only 75% of particles are controlled). Dampen down.
Respiratory protective equipment	Half face respirator or disposal face mask, rated P3 or FFP3.

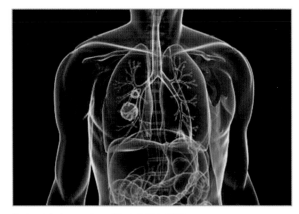

Damage to lungs caused by inhaling silica

Exposure to RCS over a long period can cause fibrosis (hardening or scarring) of the lung tissue with a consequent loss of lung function. Sufferers are likely to have severe shortness of breath and may find it difficult or impossible to walk even short distances or upstairs. The effect continues to develop after exposure has stopped and is irreversible. Sufferers usually become house- or bed-bound and often die prematurely due to heart failure.

Silicosis

Silicosis is a huge problem within the construction industry. Respirable crystalline silica (RCS) is found in stone, rocks, sands and clays, and so is contained in many construction products (such as concrete, blocks, bricks, ceramics, and so on). Common day-to-day activities produce RCS and many workers are exposed to it knowingly and unknowingly by breathing in silica particles.

Dust and fumes (Respiratory risks)

Acute silicosis is a rare complication of short-term exposure to very large amounts of silica. This condition is life-threatening and associated with significant clinical consequences.

Silica may also be linked to lung cancer. Precautions taken to control the risk of fibrosis will serve to control the risk of lung cancer. Workers with silicosis are at an increased risk of tuberculosis, kidney disease and arthritis. Exposure to RCS may also cause chronic obstructive pulmonary disease (COPD).

Cement boarding used as part of a vertical cladding system had to be cut to size on site. As can be seen in the photo below, a circular saw complete with vacuum collection was used in a designated cutting area and the operator also wore FFP3-rated RPE.

Rubbing down tape and jointing. Traditionally rubbing down creates a lot of dust, which is hazardous to health, has to be cleaned up and is manually intensive. This drywall sander with collection system minimises all these issues to the extent that the operative does not need RPE.

Dust and fumes (Respiratory risks)

B 10

11

Noise and vibration

What your employer should do for you	164
What you should do as a supervisor	165
Noise	166
Vibration	170

Noise and vibration

What your employer should do for you

1. Establish whether noise is to be either eliminated or controlled.
2. Implement systems to carry out noise assessments to ensure exposure is controlled and does not exceed the set limits.
3. Provide adequately signed ear protection zones.
4. Make hearing protection available and provide training in its use.
5. Provide vibration exposure monitoring and information for at-risk workers.
6. Provide low vibration work equipment where possible.
7. Establish routine medical surveillance for at-risk workers.

Noise and vibration

What you should do as a supervisor

Checklist	Yes	No	N/A
1. Co-operate with the employer to ensure noise is either eliminated or controlled.			
2. Monitor systems that carry out noise assessments to ensure exposure is controlled and does not exceed the set limits.			
3. Check that ear protection zones are adequately signed and rectify faults.			
4. Ensure hearing protection is available and worn and assist with training in its use.			
5. Explain vibration exposure monitoring to at-risk workers, seek clarification where uncertain.			
6. Ensure that low vibration work equipment is used where possible.			
7. Ensure routine medical surveillance for at-risk workers is carried out.			

B11

Noise and vibration

Noise

Noise is often taken for granted as an everyday feature of working on construction sites. Many construction activities create levels of noise that have the potential to cause permanent damage to the hearing of those affected if protective measures are not taken.

In addition to the long-term health implications, noise can also affect safety on site by interfering with spoken communications to the point where the risk of accidents is increased.

As well as hearing problems, excess noise is also known to cause:

- annoyance and irritation
- loss of concentration
- reduced efficiency
- fatigue
- headaches
- increased accident-proneness
- masking of warning signals.

There is a lot of equipment that produces high levels of noise in the building and construction industry. Sound levels are measured in decibels (dB) and some typical examples include:

- Electrical hand tool — 95 dB
- Hammer drill — 102 dB
- Circular bench saw — 107 dB
- Rock drill — 115 dB
- Bulldozer/grader — 121 dB

	SOUND INTENSITY RATIO	SOUND LEVEL IN dB(A)	SOUND SOURCE PAIN THRESHOLD
Harmful range	100 000 000 000 000	140	Jet engine
	10 000 000 000 000	130	Riveting hammer
			THRESHOLD OF FEELING
Critical zone	1 000 000 000 000	120	Propeller aircraft
	100 000 000 000	110	Rock drill
	10 000 000 000	100	Plate fabrication shop
	1 000 000 000	90	Heavy vehicle
Safe range	100 000 000	80	Very busy traffic
	10 000 000	70	Private car
	1 000 000	60	Ordinary conversation
	100 000	50	Quiet office
	10 000	40	Soft music from radio
	1 000	30	Quiet whisper
	100	20	Quiet urban dwelling
	10	10	Rustle of a leaf
	1	0	THRESHOLD OF HEARING

Typical sound intensities

Noise is measured on a logarithmic scale, which means that an increase of 3 dB results in a doubling of the noise level. (For example, 87 dB is twice the noise level of 84 dB.)

Noise legislation

The Control of Noise at Work Regulations (CNAWR) sets out exposure action levels of noise to which employees may be exposed as follows:

- a lower exposure action value of 80 dB(A) or peak sound pressure of 135 dB(C)
- an upper exposure action value of 85 dB(A) or peak sound pressure of 137 dB(C)
- an exposure limit value of 87 dB(A) or peak sound pressure of 140 dB(C).

Noise and vibration

Exposure limit value 87 dB(A)

The exposure limit value is the maximum level of noise to which anyone may be exposed, as measured at the ear (for example, inside ear defenders). This means that:

- ☑ ambient noise levels can exceed that value providing that any control measures used (including personal protective equipment (PPE)) limit the level of noise sensed by any individual to below that limit
- ☑ employers have a responsibility to take the performance capabilities of PPE into consideration when purchasing it and when carrying out risk assessments.

 Employers have a duty to reduce the risk of damage to employees' hearing to the lowest level that is reasonably practicable.

Lower exposure action level – 80 dB(A)

At this level of exposure employers must:

- ☑ assess the risks to the hearing of employees from noise and identify the measures necessary to comply with CNAWR. (this may involve having a noise survey carried out)
- ☑ make ear protection available to anyone who requests it
- ☑ make sure that it is maintained in good condition
- ☑ make employees aware of:
  - the dangers of exposure to noise
  - what they must do to protect themselves
  - how and where ear protectors can be obtained
  - their legal duties under these regulations.

Upper exposure action level – 85 dB(A) and above

At this level of exposure employers must:

- ☑ reduce noise levels to as low as is reasonably practicable, other than by the provision of ear protectors
- ☑ identify areas of the site in which employees will be subjected to this level of noise and display signs (as shown below) in appropriate places to indicate hearing protection zones
- ☑ provide health surveillance and supply appropriate hearing protection devices, ensuring that these devices are properly used by employees at all times that they are in the hearing protection zone.

Employees must use hearing protection as directed and report any loss or defects to their employer.

Noise and vibration

The measurement of noise

The measurement of noise is a specialist activity. An individual's exposure depends not only on the level of noise, which will usually vary as different activities start and stop, but also on the length of time that they are exposed to noise.

Advice on estimating the level of noise can be found at www.hse.gov.uk/noise

The various action values are meaningless to anyone who does not have a way of accurately calculating the noise levels on site, taking into account the time factor.

There are, however, two rules of thumb that should give a rough indication of noise levels.

- ✓ If two people standing 1 m apart (about an arm's length), who are not wearing hearing protection, have to raise their voices to hear each other, the noise level is around or above 90 dB. Hearing protection must be worn to reduce the noise at the ear to below the exposure limit (ELV) or the noise reduced to the ELV by other means. Employers must identify why ELV has been breached and modify organisational and technical measures.

- ✓ If two people standing 2 m apart (double arm's length) have to raise their voices to hear each other, the noise level is around 85 dB and hearing protection must be worn.

There is also a third rule that, if there is any doubt, everyone affected by the noise should be wearing hearing protection in the interests of their health.

Where there is likely to be a persistent noise problem, employers should arrange for noise surveys to be carried out across the whole scope of their operations. These surveys must only be carried out by a competent person who has received proper training in noise measurement techniques and has the appropriate noise measuring equipment.

There are some quite simple control measures that employers can put into place to either reduce the level of noise created and/or reduce the exposure of employees to the noise that is created.

The control of noise

The most effective way of controlling exposure is to eliminate the source of noise, thereby removing the risk of damage to hearing. This is often not practical although sometimes a change of work process is possible.

Where noise elimination is not practical, there are several control techniques that might be applied. Which method is chosen will depend upon the nature of the equipment, the way in which it is used and the work environment.

Control methods include:

- ✓ isolating a sound source from the area around it (for example, a measure as simple as positioning a mobile compressor behind stacked materials)

- ✓ insulating noisy equipment by enclosing it in sound-reducing materials

- ✓ absorbing sound with acoustic screens, or similar

- ✓ damping of sound by mounting plant and equipment on rubber or other similar sound-absorbing substances

- ✓ silencing (for example, maintaining exhaust systems on internal combustion-driven equipment in good order)

- ✓ maintaining mechanical equipment in a good state of repair.

Noise and vibration

 It is only when the use of other control measures has been explored and found to be not reasonably practicable that the use of PPE (such as hearing protectors) can be considered as the last resort.

Selection of hearing protection

In order to select the correct type of hearing protection you need to know the noise level on site and be able to select the right type of ear protection required to reduce the noise exposure to a safe level (for example 100 dB(A) would need to be reduced by 20 to bring it to 80 dB(A)). This reduction is known as attenuation and this value is shown on manufacturers' products.

The HSE recommends providing hearing protectors that perform 4 dB better than the required minimum, in order to take into account real world factors (such as poor fitting).

This means that to protect the worker against 100 dB(A) of noise you need to select hearing protection with attenuation of 24 dB(A).

The two basic types of hearing protection are earplugs and ear defenders.

Disposable earplugs. These are made from a range of materials, and are sometimes ready shaped. They must be inserted correctly following the manufacturer's instructions. Some may be reused until they no longer fit. They should only be handled with clean hands.

Reusable earplugs. These are made of rubber or plastic, and need regular and careful washing. Initial supply and fitting must be by a trained person and different sizes may be required for each ear. They must be a good fit. Dirt from unclean equipment can cause ear irritation and possible infection.

Ear defenders. These comprise a headband with protective shells that completely cover the ear. The shells have foam or liquid-filled seals that exclude noise from the ears. Poorly designed or old ear defenders may give little or no protection.

- ☑ Loose ear defenders are ineffective and mean loss of protection.
- ☑ Worn or damaged seals also mean loss of protection.
- ☑ Suitability and fitting must take account of whether spectacles are worn.
- ☑ Ear defenders fitted to safety helmets must fit snugly to the ears and not move with the helmet. (Some employers do not permit the use of helmet-mounted ear defenders because of a (generally) poor fit.)
- ☑ Ear defenders must only be used for the noise conditions they were designed for. They may not prove effective in other conditions.

Ear protection is only effective while being used. If protection is worn for only half the shift, only some 10% protection is gained. If worn for seven hours out of eight, the protection factor is still only 75%.

Continued exposure to loud or sudden noise can cause deafness.

 Use your ear protection equipment. Do not become a victim.

 For further information on the provision and use of hearing protectors refer to GE 700 *Construction site safety,* Chapter B13 Noise.

Noise and vibration

Vibration

Exposure to excessive levels of vibration has been known to be a cause of occupational ill health for many years and employers were expected to manage the situation with little practical guidance. The Control of Vibration at Work Regulations put in place legislative requirements with which employers have to comply if their employees are exposed to harmful levels of hand-arm or whole-body vibration.

Hand-arm vibration

The most common form of vibration affecting those who work in construction is hand-arm vibration syndrome (HAVS), which has the potential to damage the circulation, nerves, joints and bones in the hands and arms. HAVS is usually caused by the continued use of rotating or percussive tools (such as disc cutters, needle guns, hammer-action drills, hand-held sanders, brush-cutters, chainsaws, and so on).

The factors that can influence the degree of severity of HAVS include:

- ☑ the frequency at which the tool vibrates
- ☑ the exposure pattern (length and frequency of work and subsequent rest periods)
- ☑ the sharpness and suitability of the cutting tool or blade to cut the material being worked on
- ☑ the grip, push and other forces used to guide and apply the vibrating tools; the tighter the grip, the more vibration energy is transferred to the hands (vibration may increase as a tool becomes blunt and the user pushes harder)
- ☑ other factors that potentially affect blood circulation (such as workplace temperature, whether the person smokes and individual susceptibility)
- ☑ the hardness of the material being worked
- ☑ the posture of the tool user, because tense muscles are more susceptible to hand-arm vibration (for example, using vibrating tools at arm's length increases the effect).

The symptoms

Hand-arm vibration syndrome can develop into a severe and potentially disabling condition known as vibration white finger.

Symptoms of vibration white finger are usually triggered by the cold. Early signs are:

- ☑ pale fingertips
- ☑ loss of sense of touch
- ☑ severe pain and numbness
- ☑ pins and needles
- ☑ loss of grip
- ☑ loss of dexterity (doing up buttons, and so on)
- ☑ in the longer term, vibration white finger can result in disabling conditions, including ulceration and gangrene.

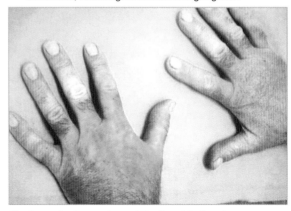

Hand-arm vibration syndrome (vibration white finger)

Noise and vibration

Managing hand-arm vibration

The ideal way of managing hand-arm vibration is to eliminate it altogether by doing jobs another way (such as using remotely controlled tools). However, this is often not possible or practical. Where hand-arm vibration cannot be eliminated and may exceed the exposure action value, the employer must introduce a programme of organisational and technical measures (control measures) consistent with the findings of a risk assessment. This must include consideration of:

- ☑ adopting a buy smooth policy, in which a commitment is made to replace old tools with new ones that incorporate low vibration technology
- ☑ the proper maintenance of tools that cause vibration
- ☑ suitable and sufficient information and training for employees so that they may use the equipment safely and correctly in order to minimise their exposure to vibration
- ☑ the reduction of the exposure period for the operative through job rotation
- ☑ the provision of clothing to protect employees from cold and damp
- ☑ the facility to regularly check and replace worn and blunt drill bits, blades, points and chisels
- ☑ advising at-risk employees of the importance of keeping the muscles warm and adopting a good posture
- ☑ providing heath surveillance for any employees thought to be at risk.

HAVS is preventable, but once the damage is done it is permanent.

Many hire companies have a traffic light system for vibrating tools that grades each tool 'green', 'amber' or 'red'. Guidance is provided on how long each type of tool should be used to avoid vibration-induced health problems.

It is worth noting that these time limits are actual total trigger time or machine running time – not how long the job takes.

Hand-arm vibration syndrome is a notifiable disease under the Reporting of Injuries, Diseases and Dangerous Occurrences Regulations (Revised) (RIDDOR) and must therefore be reported to someone in authority within the company. If in any doubt, contact your local HSE office or the HSE Employment Medical Advisory Service (EMAS).

Whole body vibration

A less common form of vibration is whole body vibration (WBV) that can affect the body when vibration is transmitted to all of the body, through a supporting surface (for example, through the seat of ride-on plant). Excessive exposure to WBV has been linked with long-term back pain.

Modern ride-on plant should have seats and, in some cases, cabs that are sufficiently sprung or cushioned to reduce the chance of WBV becoming a problem. Employers will need to be aware of the potential problem if plant operators are using older equipment that results in jarring of the back. In such circumstances the level of vibration and the length of time of exposure will be factors that determine an individual's exposure. Limiting the exposure of operators by job rotation may be the most practical solution in some circumstances.

Noise and vibration

Vibration is less of an issue in modern plant

Measuring vibration

The measurement of vibration is usually carried out by specialist companies, although some larger construction companies have bought their own vibration measuring equipment and trained their staff to carry out in-house vibration measurements. The legislation specifies two levels of vibration that require employers to take action:

- ☑ exposure action value is a level of vibration that if reached, as determined by vibration measurements, requires that employers put in place control measures to manage the exposure of employees

- ☑ exposure limit value is a measure of vibration, again determined by measuring vibration, above which employees may not be exposed and therefore the activity must stop.

The exposure action and limit values are shown below.

Hand-arm vibration	
Daily exposure action value	2.5 metres per second squared (2.5 m/s²A(8)).
Daily exposure limit value	5 metres per second squared (5 m/s²A(8)).
Whole body vibration	
Daily exposure action value	0.5 metres per second squared (0.5 m/s²A(8)).
Daily exposure limit value	1.15 metres per second squared (1.15 m/s²A(8)).

Note: the A(8) notation after each value indicates that the measurement of exposure to vibration is time weighted over an eight-hour period.

So a machine with a rating of 2.5 m/s² means it has a daily exposure action value of eight hours, therefore it can be safely used for eight hours.

Other examples are:

- ☑ 5.0 m/s² means it can be safely used for 2 hours
- ☑ 7.5 m/s² means it can be safely used for 53 minutes
- ☑ 10 m/s² means it can be safely used for 30 minutes
- ☑ 15 m/s² means it can be safely used for 13 minutes
- ☑ 20 m/s² means it can be safely used for 8 minutes.

Thus machine selection and their vibration ratings greatly affect the amount of time equipment can be used safely.

Noise and vibration

 The HSE's vibration calculator can be found at www.hse.gov.uk/vibration/hav/vibrationcalc.htm

Multiple tool use during the day must also be taken into account, especially if they have different vibration ratings and are used for varying times.

 For further information refer to GE 700 *Construction site safety,* Chapter B14 Vibration.

Medical surveillance

Where employees' exposure to noise or vibration is likely to exceed action levels, as defined in the relevant regulations, the employer has a duty to arrange medical surveillance for the affected employees.

 For further information on medical surveillance refer to Chapter B08 Health and welfare.

Noise and vibration

12

Hazardous substances

What your employer should do for you	176
What you should do as a supervisor	177
Introduction	178
Control of Substances Hazardous to Health Regulations	178
Identifying hazardous substances	180
The COSHH assessment	181
REACH Regulations	183
Control of Lead at Work Regulations	183

Hazardous substances

What your employer should do for you

1. Provide hazard data sheets for each product and a COSHH assessment written for each where necessary.
2. Take appropriate measures to prevent or control exposure to hazardous substances, including those created by work processes and/or those already on site.
3. Monitor the effectiveness of control measures to ensure they are adequate.
4. Ensure that the hazard data sheet and COSHH register is readily available to the operatives likely to be exposed.
5. Issue the correct personal protective equipment (PPE) and respiratory protective equipment (RPE) with instructions, signed for by the work team.
6. Provide information on the hazardous substances used for inclusion in the health and safety file.
7. Establish correct emergency control measures and ensure they are in place and provide kits to be available in case of incident.
8. Provide health surveillance where it is necessary.

Hazardous substances

What you should do as a supervisor

Checklist	Yes	No	N/A
1. Understand the contents of hazard data sheets and COSHH assessments prior to work and explain control measures to the work team.			
2. Use appropriate measures to prevent or control exposure to hazardous substances, including those created by work processes and/or those already on site.			
3. Report any shortfall in the effectiveness of control measures to ensure they are adequate.			
4. Check that the hazard data sheet and COSHH register is readily available to the operatives likely to be exposed.			
5. Issue the correct PPE and RPE and provide instructions and understanding in the correct use by the work team.			
6. Gather any information on the hazardous substances to be used and include in the health and safety file for use by your employer.			
7. Understand the correct emergency control measures and ensure they are in place and that kits are available in case of incident.			
8. Submit yourself and your work team for health surveillance where it is necessary.			

B 12

Hazardous substances

Introduction

Many substances that are either used or are created as a by-product of a work activity have the potential to harm the health of the person(s) involved in the activity and others who are near enough to be affected.

It will be necessary for your employer to carry out a risk assessment to:

- identify the hazardous substances used in, or created by, a work process to which employees and others will be exposed
- establish the degree of risk to their health resulting from exposure
- devise safe systems of work that either eliminate exposure or control it to an acceptable level.

As with controlling any risk, the most effective course of action is to avoid using or creating the hazardous substance in the first instance. This chapter will outline the requirements of:

- the Control of Substances Hazardous to Health Regulations
- Regulatory framework for management of chemicals (REACH)
- the Control of Lead at Work Regulations.

Control of Substances Hazardous to Health Regulations

These regulations, usually referred to as COSHH, place legal duties on employers, employees and the self-employed. They are aimed at the protection of employees and others from the effects of working with any substances that can result in health problems. These substances are not limited to those that are obtained from suppliers but also include some that are created by the work process.

The regulations do not cover exposure to asbestos, lead or radioactive substances, which have their own regulations. *(Lead is covered separately in this chapter and asbestos is in B10 Dust and fumes (Respiratory risks).)*

The objective of the regulations is to prevent, as far as possible, the exposure of employees to hazardous substances and, when this cannot be achieved, to control the level of exposure that does occur.

The objective is usually met by carrying out a (COSHH) risk assessment and implementing the necessary measures identified.

Many of the substances commonly used in construction activities have the potential to cause serious health problems if the risks are not evaluated and exposure not properly controlled.

 The Health and Safety Executive (HSE) publishes a guidance note *Occupational exposure limits* (EH40), which lists the range of substances where limits to exposure have been set and their exposure limits.

 The online version is updated periodically, at www.hse.gov.uk/pubns/books/eh40.htm

Hazardous substances

The workplace exposure limit (WEL), given for a wide range of substances, must be complied with if work is to be carried out safely and without risks to health and safety of the workers.

Hazardous substances commonly found on construction sites are usually in the form of dust, fumes, vapour, gas and mists and include:

- cement, plaster and the dust of some woods
- paint, thinners, bitumen and other sealants
- solvents, adhesives and acids
- vehicle fuels, hydraulic and other oils.

Operatives priming a slab, wearing PPE as identified in their risk assessment

Hazardous substances that may be created by a work process or are already on site include:

- contaminants in soil (brownfield sites)
- silica and other types of dust from cutting or cleaning operations
- fumes from hot-work processes
- an accumulation of bird droppings
- naturally-occurring gases in confined spaces.

Hazardous substances may be:

- inhaled
- swallowed
- absorbed through the skin (broken or unbroken)
- introduced to the body via a puncture of the skin.

Depending upon the nature of the substance, the effects may be instant or may only become apparent after several years. They can result in long-term or permanent illness and, in some cases, death.

 Even natural materials can be harmful. For example, henna can cause dermatitis and asthma, wood dust can cause asthma, stone or concrete dust can cause lung disease such as silicosis, and citrus oils can cause skin irritation.

Identifying hazardous substances

The packaging or container of hazardous substances will carry one or more of the symbols below, to indicate its hazardous properties. Since 2009, new international symbols have been gradually replacing the European symbols. Both are shown in the following table.

New international symbols		Existing European hazard (CHIP) symbols	
	Fatal if inhaled, swallowed or in contact with skin.		**Very toxic or toxic** Substances that, in very low quantities or low quantities, cause death or acute or chronic damage to health when inhaled, swallowed or absorbed via the skin.
	Warning – harmful if inhaled, swallowed or in contact with skin. Can also cause serious eye irritation.		**Harmful** Substances that may cause death or acute or chronic damage to health when inhaled, swallowed or absorbed via the skin.
	Warning – harmful if inhaled, swallowed or in contact with skin. Can also cause serious eye irritation.		**Irritant** Non-corrosive substances that may cause inflammation through immediate, prolonged or repeated contact with the skin or mucous membrane.
	Danger – causes severe skin burns and eye damage.		**Corrosive** Substances that may, on contact with living tissues, destroy them.
	Danger (Category 1) – may cause allergy or asthma symptoms or breathing difficulties if inhaled. Danger (Category 1a and 1b) – may cause cancer.	None	

Hazardous substances

The suppliers of all such substances have a legal duty to also supply a safety data sheet for each product from which the COSHH assessment may be compiled.

Of course, hazardous substances that are created by a work process will carry no such health warning. These must be identified by the findings of a risk assessment.

As in all risk assessments, the effort put into compiling it and the degree of control measures necessary need only be proportional to the degree of risk.

The COSHH assessment

Employers have the duty to make an assessment, but they can arrange for anyone who is competent to do so to assist with the task, whether from within the company or from outside.

It is the responsibility of the employer to ensure that anyone who carries out any duties under the COSHH Regulations is competent to do so, and has received the necessary information, instruction and training, whether or not the assessor is an employee.

The majority of assessments for small building and construction work can be carried out without the need for the assistance of specialist outside consultants, provided the assessor is competent and:

- ☑ has access to the safety data sheets for the products concerned
- ☑ understands, in basic terms, what the COSHH Regulations require to be done
- ☑ is knowledgeable of the work process being assessed
- ☑ has the ability to systematically gather relevant information regarding exposures to hazardous substances and the subsequent risks to staff by:
  - observing working practices
  - obtaining information on substances used
  - asking questions in the workplace
  - making informed 'what if' judgements regarding possible divergences from standard working practices
- ☑ can specify the steps and control measures to be taken to comply with the COSHH Regulations
- ☑ can appreciate their own limitations; knowing when to call in specialists with certain skills (such as when there is a need to undertake air sampling)
- ☑ has the ability to make valid conclusions, to make a report and communicate findings regarding risks and precautions to the employer and employees.

The main requirements of the regulations are as follows.

Know what products and substances you are using

- ☑ Compile a list of the hazardous substances that are in use.
- ☑ Add to the list any hazardous substances that are created as a by-product of a work process.

Assess the health hazards they can cause

- ☑ From manufacturer's information and other relevant sources, determine the level of risk to health, the degree of exposure and what action is needed to eliminate or control exposure.
- ☑ Carry out a COSHH risk assessment to establish how the hazardous substances might cause harm. Record the findings of the risk assessment and the control measures to be taken.
- ☑ Review any assessment regularly or whenever the exposure monitoring results indicate that it is necessary.

B 12

Hazardous substances

Eliminate or control the risk of exposure by:

- ☑ designing a work process that prevents exposure to, or the creation of, hazardous substances (this is not possible in many construction activities)
- ☑ substituting or diluting the substance (designers and specifiers should look towards less hazardous options)
- ☑ using dust or fume extraction (this may be a standalone extractor unit or built-in to some hand tools)
- ☑ using engineering controls, such as totally enclosing the process (often not feasible in the construction environment)
- ☑ issuing suitable personal protective equipment (PPE) including respiratory protective equipment (RPE) and making sure it is worn (only after the use of all other risk control measures have been explored).

Before work starts, give information, instruction and training to employees relating to:

- ☑ the nature and degree of known risks
- ☑ the control measures adopted and how they should be put into operation
- ☑ reasons for, and the correct use of, PPE
- ☑ any exposure monitoring arrangements that are in use
- ☑ the purpose of, and arrangements for, any health surveillance, if appropriate.

Issuing personal protective equipment

Obtaining appropriate PPE for hazardous substances that can affect the skin might be as simple as identifying the correct type of gloves. However, where there is need for respiratory protection, selecting the correct type of RPE is critical.

 A filtering face mask designed to filter out dust will be useless against gases, fumes or vapours.

The need for employees to work in masks or respirators is always the last resort. If RPE is necessary, it must:

- ☑ adequately control exposure to the hazardous substance(s) identified
- ☑ suit the wearer (comfort and fit)
- ☑ be appropriate for the job
- ☑ be used correctly.

Before selecting RPE, proper thought must also be given to:

- ☑ the physical condition of the employee; breathing through some types of RPE for extended periods can require effort
- ☑ whether facial hair or glasses would make the equipment ineffective
- ☑ providing suitable training on the safe use and maintenance of the RPE.

Health surveillance

In many cases, health surveillance must be carried out by an occupational health practitioner and recorded. Employees must have access to medical records that apply to them.

Health surveillance has to be undertaken when an employee is exposed to:

- ☑ one of the substances and is engaged in a process listed in Schedule 6 of the COSHH Regulations, to which reference should be made, although those listed are unlikely to apply to the construction industry

Hazardous substances

- ☑ a hazardous substance that is linked to an identifiable disease related to the exposure (for example, exposure is known to cause asthma) and there are valid techniques for detecting indications of the disease.

Monitor the effectiveness of any controls

Are the measures that are in place to control exposure effective?

Observation of the task and speaking to the people doing the job will usually provide the answer.

This may also require the regular testing of equipment (such as ventilation), which must be kept in efficient working order.

Keep records

Monitoring arrangements and health surveillance must be recorded. Some records may need to be kept for 40 years.

In companies where there are union-appointed safety representatives, or representatives of employee safety, the above information should be available to them.

Produce accident, incident and emergency plans, including:

- ☑ making appropriate first aid provision
- ☑ developing and practicing safety drills (for example, site evacuations), as appropriate
- ☑ identification and details of hazards
- ☑ making available details of specific hazards that are known to exist
- ☑ developing warning and communication systems
- ☑ displaying emergency plans
- ☑ making relevant information available to the emergency services.

For further information refer to GE 700 *Construction site safety,* Chapter B07 The Control of Substances Hazardous to Health.

REACH Regulations

REACH is a European Community Regulation on chemicals and their safe use (EC 1907/2006). It deals with the **R**egistration, **E**valuation, **A**uthorisation and **R**estriction of **CH**emical substances.

This regulatory regime for registering chemicals, which came into effect on 1 June 2007 and will be fully implemented over the next ten years, will have some impact on the way in which information is made available since it requires suppliers to consider the possible effect that chemicals will have in use.

This is very different to the present situation where the user, who may know very little about the chemicals, has to make the assessment. REACH will put the responsibility onto those who should be in the best position to do this. It is likely that the contents of material safety data sheets will have to change as a result of these regulations.

Control of Lead at Work Regulations

Those people who actually work with lead are most at risk of suffering ill health from exposure to it. However, other people who are working close by may also be at risk, depending upon the nature of the job being carried out and how well the risks of exposure are controlled.

Hazardous substances

Lead is a cumulative poison that collects in the bone marrow and affects the body's capability to produce new blood cells. It is usually taken into the body in the form of dust or fumes, by ingestion (via the mouth or through the skin) or breathed in.

The regulations aim to give greater protection of health to people at work by reducing their exposure to lead and thus the concentrations of lead in their blood. Where concentrations prove to be too high, employers are required to remove employees from work with lead. This is called the suspension level. If employers cannot transfer employees to other work not involving exposure to lead, they must pay them suspension pay under the Employment Rights Act.

Blood levels below the suspension levels, known as action levels, have been set. If these lower levels are breached, employers have a duty to investigate and remedy the cause. Employers are also required to take positive steps to reduce the concentrations of lead in air to a level not exceeding the occupational exposure limits in the regulations. Women of child-bearing age and young people have lower blood-lead suspension levels than those applying to other workers.

The Control of Substances Hazardous to Health Regulations do not apply in any situation where the Control of Lead at Work Regulations apply.

The requirements of the regulations

The main requirements of the Control of Lead at Work Regulations are shown below.

- ☑ Employers whose employees are to work with lead must carry out a suitable and sufficient assessment of the risk to the health of those employees arising from that work.

- ☑ Every employer shall ensure that the exposure of employees to lead is either prevented or, where this is not reasonably practicable, adequately controlled by means of appropriate control measures. Adequate control of exposure to lead covers all routes of possible exposure, in other words, inhalation, absorption through the skin and ingestion.

- ☑ Adequate steps must be taken to control ingestion. An employer must ensure that, as far as is reasonably practicable, employees do not eat, drink or smoke in any place which is, or is liable to become, contaminated by lead. Employees should be warned against doing so. Employers have additional duties under the Workplace (Health, Safety and Welfare) Regulations to provide suitable and sufficient rest facilities and facilities to eat meals, where food eaten in the workplace would otherwise be likely to become contaminated.

- ☑ All control measures which have been provided should perform as originally intended and be effective in protecting people from lead. Any defect in the equipment, or failure to use and apply it properly, which could result in a loss of efficiency or effectiveness, thus reducing the level of protection, should be identified and rectified as soon as possible.

- ☑ Where employees are liable to receive significant exposure to lead, employers must establish monitoring by both air sampling and measuring the concentrations of lead in both blood and urine.

- ☑ Where exposure to lead is deemed to be significant, the employer should make sure that the employee is under medical surveillance by either a medical inspector (Employment Medical Adviser) or a relevant doctor.

- ☑ Employers who undertake work liable to expose employees to lead shall provide such information, instruction and training as is suitable and sufficient to know the risks to health, and the precautions that should be taken.

Hazardous substances

Note: in addition to this, the Code of Practice that supports the legislation requires that an employer should issue a copy of the HSE's free leaflet entitled 'Lead and you' to all employees on their first employment on work with lead, and make copies available for issue at the request of any employees or their representatives. Employers must additionally provide employees with written records of concentrations of lead in air to which they have been exposed, the results of the measurements of lead in their blood or urine, and an explanation of the significance of these results.

- ☑ The employer, in an attempt to protect the health of employees from an accident, incident or emergency must ensure that procedures, including the provision of first-aid facilities and safety drills, have been prepared and can be put into effect should such an occasion arise. The employer must also ensure that information on such emergency arrangements have been notified to accident and emergency services and that all such information is displayed within the workplace.

For further information refer to GE 700 *Construction site safety*, Chapter B12 Lead.

Risk assessment

Employers must not carry out any work that may expose employees to lead unless a suitable and sufficient risk assessment has been carried out. The purpose of the risk assessment is to enable the employer to:

- ☑ assess whether the exposure of employees to lead is likely to be significant

- ☑ identify the measures necessary to prevent or control exposure.

This includes other people who are not employees but who may be exposed as a result of the way the employer carries out the work concerned. The assessment must be reviewed as often as necessary and in other certain specified circumstances, and a record made of any significant findings if more than five are employed. Such an assessment will allow the employer to make a decision on whether the work concerned is likely to result in an employee being significantly exposed to lead and to identify the measures needed to prevent or adequately control exposure.

Medical surveillance

Where employees' exposure to lead is likely to exceed action levels, as defined in the relevant regulations, the employer has a duty to arrange medical surveillance for the affected employees.

For further information on medical surveillance refer to Chapter B08 Health and welfare.

Hazardous substances

B 12

13
Manual handling

What your employer should do for you	188
What you should do as a supervisor	189
Introduction	190
Duties of the employer	190
The manual handling assessment	191
Avoiding manual handling – some solutions	192
Manual handling injuries	193
Manual handling – a few techniques	193
Team lifting	195
Using simple mechanical aids	195

Manual handling

What your employer should do for you

1. Arrange that the first consideration is to avoid manual handling where a risk has been identified.
2. Organise that delivered loads are deposited and stacked safely.
3. Ensure that loads have been assessed as to suitability for manual handling.
4. Provide mechanical aids and provide training in the use of them.
5. Take into account the elements of the task, load, environment and individual capability (TILE), including repetitive lifting and carrying.
6. Provide training in proper (kinetic) lifting and team lifting methods to the workforce and ensure that they understand that they should assess the load and if in doubt ask for help.
7. Ensure that personal protective equipment (PPE) is issued to protect hands, feet and so on.
8. Provide and maintain safe routes for carrying loads.

Manual handling

What you should do as a supervisor

Checklist		Yes	No	N/A
1.	Organise that the first consideration is to avoid manual handling where a risk has been identified.			
2.	Supervise to make sure that delivered loads are deposited and stacked safely.			
3.	Check that loads have been assessed as to suitability for manual handling.			
4.	Ensure the use of mechanical aids and check that training has been provided in the use of them.			
5.	Supervise and manage working practices that avoid repetitive lifting and carrying.			
6.	Ensure training is provided in proper (kinetic) lifting and team lifting methods to the workforce and ensure that they understand that they should assess the load and if in doubt ask for help.			
7.	Organise the issue of correct PPE and ensure that it is used correctly.			
8.	Monitor and maintain safe routes for carrying loads.			

B 13

Manual handling

Introduction

Manual handling is one of the most common causes of injury at work and is responsible for more than a third of all workplace injuries. Many cause absence from work, and in the worst cases, permanent disability and physical impairment is often the result.

The Manual Handling Operations Regulations place legal duties and responsibilities on both employers and employees to ensure that manual handling activities are planned and carried out so that injury is avoided.

Risks arising from manual handling should be minimised

The Management of Health and Safety at Work Regulations require employers to make a suitable and sufficient assessment of the risks to health and safety of their employees. This requirement is expanded upon in the Manual Handling Operations Regulations, which require employers to assess the risks to the health of their employees arising out of manual handling activities. This is, in effect, the same assessment; there is no requirement for two assessments.

 The Health and Safety Executive (HSE) has produced a comprehensive manual handling assessment chart (MAC), which can be downloaded from www.hse.gov.uk/msd/mac/index.htm

Employers should find the MAC useful in identifying high risk manual handling operations and helping them to complete their risk assessments.

Duties of the employer

 Manual handling relates to the moving of items by lifting, lowering, carrying, pushing or pulling.

The employer must identify all manual handling tasks that might involve a risk of injury and carry out a manual handling assessment of those activities. The assessment should also take into consideration the number of times the load will be lifted, the distance the load will be carried, the height it will be moved, and any twisting, bending, stretching or awkward posture which the carrier might be forced to adopt whilst undertaking the task.

Where the assessment indicates that there are risks to employees from the manual handling of loads, the employer must take steps to either:

- ☑ avoid the need for employees to carry out manual handling activities as far as is reasonably practicable, usually achieved by better planning or moving loads using mechanical means, or
- ☑ where the manual handling of loads is unavoidable, reduce the risk of injury so far as is reasonably practicable by planning how these activities can be carried out safely and putting in place suitable control measures to ensure that they are.

The manual handling assessment

An ergonomic approach to the problems of manual handling and lifting can help to overcome many of the problems.

Ergonomics have been described as 'fitting the job to the person, rather than fitting the person to the job'. This requires attention to TILE, that is the:

- ☑ **task:** what has to be achieved and by when?
- ☑ **individual:** male or female, large or small frame, age, stature, state of health/previous injuries and so on.
- ☑ **load:** is it too heavy to lift, if so, can it be broken down into smaller loads, can two people lift it?
- ☑ **(work) environment:** consider hazards en-route (such as uneven floor surfaces, slopes, steps, narrow passages and so on).

Where the assessment indicates potential risks to the health of employees from the manual handling of loads, the employer must develop a safe system of work that avoids the risks.

The employer must consider:

- ☑ **the task:**
    - can manual handling be avoided completely?
    - can the distance a load has to be moved be reduced by better on-site planning?
    - can lifting aids be used?
    - does the load have to be raised to, or lowered from, above head height?
    - does the task involve repetitive lifting?
    - is it possible to avoid lifting from the floor?
    - can the centre of gravity move (fluid loads)?
    - can it be carried close to the body?
    - does it have sharp edges?
    - should it be carried by two (or more) people?
- ☑ **the individual:**
    - do they need manual handling training?
    - do they need additional PPE?
    - is there any known reason why they might not be suitable for the job?
- ☑ **the load:**
    - can it be broken down into smaller loads?
    - is the weight known and, if not, can it be found out?
    - are there adequate handholds?
    - is it evenly balanced (centre of gravity)?
- ☑ **the environment:**
    - is the floor surface sufficiently level?
    - are there any space constraints that should be removed?

Manual handling

- is the level of lighting adequate?
- is the workplace temperature satisfactory?

Employees must co-operate by:

☑ using the appropriate equipment supplied in accordance with their training and instructions

☑ following the systems of work laid down by their employer.

 Manual handling and lifting injuries are often for life.

Avoiding manual handling – some solutions

Reducing the risk of injury can be helped by using mechanical aids such as:

☑ genie lifts
☑ kerb lifters
☑ sack trolleys
☑ pallet trucks
☑ skids
☑ telehandlers
☑ correct delivery locations
☑ built-in lifting attachments
☑ wheelbarrows
☑ stillages

☑ suction pads
☑ temporary handles/grips
☑ material hoists
☑ conveyor belts
☑ excavators/dumpers.

Hoist being used for loading out roofing materials

Placing kerbs with a mechanical suction lifter

Manual handling injuries

Injuries resulting from unsafe or incorrect manual handling can affect the:

- ✓ whole body
- ✓ back
- ✓ shoulders
- ✓ arms
- ✓ hands
- ✓ feet and toes.

Back injuries are the most common, but hernias, ruptures, sprains and strains are all conditions that can result from poor manual handling techniques.

Poor manual handling can cause permanent damage

Legislation requires that adequate and appropriate protective clothing and equipment be provided by the employer and worn by the employee. Gloves, footwear, hard hats and overalls all play an important part in reducing the type and severity of accidents arising from manual handling.

There are conflicting views on the benefits of wearing a back-support belt. Some authoritative bodies believe that wearing one may render some people more likely to suffer an injury. There is no evidence to support the theory that they reduce manual handling injury rates. There may be some benefit derived from the fact that wearing a back support will keep the lower-back muscles warm.

The cost of physical injury can be very high for the employer through:

- ✓ lost production and raised costs
- ✓ legal liabilities leading to possible prosecution.

And, for the employee, through:

- ✓ pain and suffering
- ✓ permanent disability, leading to lost wages.

Anyone who believes that they have suffered a manual handling injury, particularly a back injury, should be encouraged to seek prompt medical advice.

Manual handling – a few techniques

Plan the task

- ✓ What has to be moved?
- ✓ Where to, where from, how far and is the route clear of obstructions?
- ✓ Is it safe for one person to do it alone?
- ✓ Will help be required? If so, how much and for what purpose?

Use your body wisely

- ✓ Let the leg and thigh muscles to do the work. They cope better than the back muscles.

Manual handling

- ☑ Try to keep your spine straight – not necessarily vertical, but straight. However, slight bending of the back, knees and hips is permitted if necessary and is certainly preferable to fully flexing the back (stooping) or fully flexing the hips and knees (deep squatting).

- ☑ Once you have started to lift (taken the full weight) don't flex your back any further, which can happen if you begin to straighten before you have raised the load.

- ☑ Avoid twisting the back or leaning sideways, especially whilst the back is bent.

- ☑ Use the movement of your own body weight to get things moving. Do not snatch the load.

Bend your knees
Feet slightly apart; one foot slightly forward; balance; keep the back straight. Avoid tight clothing that prevents you from bending your knees.

Get a good grip
Use your hand – not fingers. Tilt the load slightly to get a secure grip as close to the body as possible. Keep elbows tucked in.

Lift with your legs
Do not jerk or snatch. Let the thigh muscles do the bulk of the work. Lift in stages, if necessary from the ground on to a low platform.

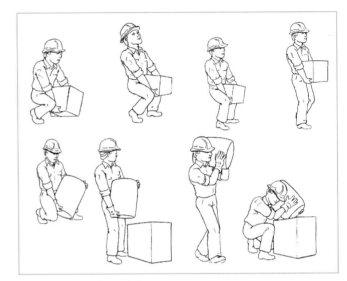

Good manual handling techniques

Putting the load down – plan before you start

- ☑ **To floor level.** It will probably be a reversal of the lifting process. Attention must be given to the positioning of the feet and back.

- ☑ **To a higher level.** Depending upon the height of the surface upon which the load is to be positioned, it may be less of a stress on the body to lower the load and assistance may be required.

Have you planned the task properly?

Manual handling

Team lifting

If the load is large, heavy or awkward – get assistance, preferably from someone of about the same size and build as yourself to help maintain the balance of the load during lifting. Always plan the lift with your helper and agree who will give directions as to when and how you will lift.

More than one person may be needed to complete a job safely

Using simple mechanical aids

Steel pipes and round timbers make effective rollers or mechanical aids, and should be used when necessary, but great care must be taken to:

 co-ordinate the movement of the load with the positioning of the rollers

 keep the hands of the person who positions the rollers well out of the way of the moving load.

The use of a wheelbarrow or sack truck will make the manual handling of suitable loads that much easier. Using lifting straps or hand-help suction devices will assist in moving some sheet materials.

The use of simple mechanical aids lowers the level of risk, can prevent accidents and avoids unnecessary fatigue and strain.

> More information can be found at
> www.hse.gov.uk/contact/faqs/manualhandling.htm

> For further information refer to GE 700 *Construction site safety,* Chapter B15 Manual handling.

Manual handling

14

Site set up and security

What your employer should do for you	198
What you should do as a supervisor	199
Legislation	200
Children trespassing on site	200
Site set up	200
Site security	201

Site set up and security

	What your employer should do for you
1.	Provide information about any known hazards near your site, including information on the surrounding area e.g. schools, transport restrictions.
2.	Provide information about any logistical issues (such as delivery times and delivery restrictions).
3.	Supply any necessary equipment that is required to provide and maintain site security, including appropriate fencing and people resources.
4.	Devise a site management plan and advise of proposed traffic management routes.
5.	Maintain up-to-date records of who is authorised to be on site at any time.
6.	Make specific security arrangements for particular pieces of plant, if required.
7.	Allow for visitor, contractor and staff parking where possible.
8.	Give consideration to the protection of nearby watercourses.
9.	Create and communicate a materials delivery schedule.

Site set up and security

What you should do as a supervisor

Checklist	Yes	No	N/A
1. Provide separate routes for pedestrians and vehicles, with separate entrances if possible.			
2. Ensure entrances to and from site minimise hazards on public roads.			
3. Ensure visitors to site are provided with safe access and know where to go.			
4. Ensure site and vehicle management plans are communicated to the workforce and their adherence monitored.			
5. Check that visitors and the workforce sign in and out as they leave site.			
6. Monitor the site security arrangements for signs of trespass.			
7. Monitor the location and routing of temporary electrical supplies.			
8. Ensure materials are stacked appropriately and hazardous materials are stored correctly.			
9. Ensure all contractors are aware of any delivery restrictions specific to the site or surrounding area.			

C14

Site set up and security

Legislation

The **Construction Design and Management Regulations** place a specific duty on the principal or main contractor to ensure that unauthorised persons do not gain access to site. Unauthorised persons will not be familiar with the site rules and site hazards, so access to site will expose them to hazards that they may not be aware of.

In accordance with the **Occupiers Liability Act**, trespassers and non employees have a right not to be put at risk if they enter a construction site. A higher standard of care is to be given by the site management to children for both authorised and unauthorised access.

Children trespassing on site

Children often think that construction sites are exciting places to play. It is therefore important that access to site is controlled during and after normal working hours. Any evidence of trespass must be reported to the site management immediately and the appropriate action should be taken. The site management team and supervisors should monitor the situation and request that operatives on site are extra vigilant and report any further incidents of trespass.

It is a good idea to visit any schools near to the site to discuss the dangers with the children and to encourage them to stay away from sites.

A video in which children warn about the dangers of playing on construction sites can be viewed at www.hse.gov.uk/campaigns/worksmart/videos/#construction

Involving school children in construction projects can help them to understand the dangers on site

Site set up

The construction phase health and safety plan provides specific information on hazards that need to be considered during the setting up phase of a construction site. The hazards could include information about the surrounding area or hazards that are present on site, for example buried services, asbestos or residual hazards from existing workplace activities. All hazards need to be identified and considered to ensure a safe site set up. The site set up needs to take into consideration any logistical issues and transport routes and restrictions.

The site office should be located near the entrance to site, where possible, and it should be accessed via a safe pedestrian route to allow for safe access for visitors and deliveries to site.

Site set up and security

Considerations for a safe site set up include:

- ☑ suitability of ground conditions for site cabins (if they are being used)
- ☑ design of the site hoardings and gates to withstand wind loadings
- ☑ material storage areas to be accessible and of sufficient size (consideration should be given to implementing height restrictions)
- ☑ plans for traffic and pedestrian routes, and safe access from the highways
- ☑ segregation and safe disposal of waste
- ☑ securing the site, preferably with lockable gates and sufficient hoarding
- ☑ location of overhead and buried services
- ☑ presence of wildlife in the surrounding area (such as bats and nesting birds)
- ☑ welfare facilities for both male and female workers
- ☑ emergency arrangements appropriate to the hazards on site
- ☑ requirements for safety signs and notices and their locations.

Security and safety considerations need to be taken into account before work begins

Site security

Unauthorised persons will probably not be aware of the hazards associated with construction sites. The site could be secured with the use of site hoarding and lockable gates. The site boundary should be secured immediately after site possession. If the whole site cannot be secured then areas with potentially hazardous operations should be appropriately secured.

A number of simple precautions can be taken to ensure that unauthorised persons cannot access the site and tools and equipment on it.

Site set up and security

- ☑ Power tools, plant and equipment should be locked and stored when not in use, particularly out of normal working hours.
- ☑ Hazardous substances should be kept in a secure designated area.
- ☑ Bottled gas compounds should be kept securely locked.
- ☑ Site accommodation should be locked out of working hours to prevent vandalism and theft of personal possessions.
- ☑ All ladders should be removed, or boarded up, to prevent access at the end of the working day.
- ☑ Plan and monitor routine checks of the site hoarding or fencing and address issues identified.
- ☑ Remove keys from plant and equipment when not in use.

15

Fire prevention and control

What your employer should do for you	204
What you should do as a supervisor	205
Introduction	206
The management of fire risks	206
Relevant legislation and its enforcement	207
Fire prevention	209
Conditions required for a fire to start	210
Hot-work permits	210
Managing fire risks	210
Emergency procedures	213
Highly flammable liquids and liquefied petroleum gases	215
Fire-fighting	217
Types of portable fire extinguisher and what to use them on	220

Fire prevention and control

What your employer should do for you

1. Establish a means of preventing the accumulation of flammable waste material and manage the additional fire risks, which arise from work processes, in addition to general fire precautions.
2. Provide and maintain an adequate number of serviceable fire extinguishers, of the appropriate type(s).
3. Train an appropriate number of site staff in the selection and use of hand-held fire extinguishers.
4. Ensure that fire extinguishers are located at well signed and easy to find fire points.
5. Install an effective fire alarm system, up-graded as work progresses.
6. Arrange for adequate emergency escape routes, with correct signage, which are large enough for the number of people who would have to use them.
7. Induct everyone on site of the fire alarm sound and what actions to take.
8. Display suitable and sufficient fire safety signs.
9. Provide suitable storage areas for flammable substances, like LPG, solvents and paints.
10. Establish smoking areas to be well-defined, safe areas of the site and provide signs and suitable bins.
11. Provide a hot-work permit system and set times for checking a location after hot work has ceased.
12. Make contact with the relevant emergency services to ensure they have sufficient information to deal with any foreseeable emergency that might arise.

Fire prevention and control

What you should do as a supervisor

Checklist	Yes	No	N/A
1. Maintain a means of preventing the accumulation of flammable waste material and check that additional fire risks, which may arise from work processes, are reported.			
2. Check fire extinguishers are serviceable and that they are at their correct locations.			
3. Monitor that the appropriate number of site staff have been trained in the selection and use of hand-held fire extinguishers.			
4. Test the fire alarm system regularly and report areas where grades are necessary as work progresses.			
5. Check emergency escape routes, with correct signage, are large enough for the number of people who would have to use them and kept clear of obstructions.			
6. Make sure that everyone on site is aware of the fire alarm sound and what actions to take.			
7. Maintain suitable storage areas for flammable substances, like LPG, solvents and paints.			
8. Check smoking areas are well-defined, observed and provided with signs and suitable bins are regularly emptied.			
9. Manage the hot-work permit system and comply with times for checking a location after hot work has ceased.			
10. Liaise with managers and check that contact with the relevant emergency services is maintained to ensure they have sufficient information to deal with any foreseeable emergency that might arise.			

C 15

Fire prevention and control

Introduction

Every year there are a number of large fires on construction sites. Many are in buildings that are undergoing alteration and refurbishment, but the move towards timber frame and system-build housing has also been the cause of several serious fires. All have serious consequences: people are injured and buildings so badly damaged that they have to be demolished. Some irreplaceable buildings burn and are lost forever.

The risk of fire is greater during the construction, refurbishment or demolition of buildings than at any other time, and the resultant loss of equipment, working time and financial implications can be severe.

Very few people actually burn to death in fires. The majority of people who lose their lives in fire situations are actually overcome by smoke and are asphyxiated. Smoke, when confined, is equally as dangerous as flame.

Serious fires are devastating

The increasing use of timber frame structures and modular building techniques have significantly increased the fire-loading on the sites where they are used. There have been several serious fires on such sites, some of them attributed to arson.

For this reason, effective site security is an absolutely necessary part of fire risk management.

The management of fire risks

Guidance contained in the 2010 edition of the Health and Safety Executive (HSE) publication, *Fire safety in construction* (HSG168), advises that effectively managing the risks of fires on construction sites must involve managing:

- ☑ **'everyday' general fire precautions (GFP), which include structural features and equipment necessary to achieve fire safety. These include ensuring that:**
  - the structure is compartmentalised at the earliest opportunity
  - emergency escape routes and exits are designated and kept clear
  - serviceable and suitable fire extinguishers are provided around the site
  - good housekeeping prevents the build-up of flammable rubbish
  - an emergency fire-action plan is drawn up
  - site security is effective to prevent arson, particularly out of normal working hours

- ☑ **the additional fire risks that arise from work processes that form a part of the job, such as:**
  - any type of hot works
  - the storage of bulk flammable substances or materials, prior to them being used

Fire prevention and control

– work on high risk projects (such as timber frame and multistorey structures), because of the high fire-loading until they can be compartmentalised and other fire-engineered solutions are installed.*

* The HSE indicates that it regards large timber frame construction and multistorey (new-build or refurbishment) jobs as particularly high risk.

Relevant legislation and its enforcement

☑ **Health and Safety Executive (HSE)** will enforce fire safety legislation on construction sites under the requirements of the Construction (Design and Management) Regulations (CDM) and the Management of Health and Safety at Work Regulations. Regulation 38 of CDM outlines the legal responsibilities with regard to the prevention of fires on construction sites; Regulations 39 and 40 deal with emergency procedures and emergency routes and exits respectively. However, readers should note that HSE inspectors also have enforcement powers under Regulatory Reform (Fire Safety) Order.

☑ **Fire and Rescue Service (FRS)** will enforce fire safety legislation under the Regulatory Reform (Fire Safety) Order RR(FS)O (in England and Wales) or the Fire (Scotland) Act, in premises other than construction sites, including for example, a construction company's office premises or off-site storage yards. The Fire Safety (Employees' Capabilities) (England) Regulations require that employers in England take into account the capabilities of employees, with regards to their knowledge and training in fire safety matters before allocating fire safety responsibilities to them. However, the local Fire and Rescue Service (FRS) do have enforcement powers on some construction sites as outlined in the chart on the next page.

The legal requirements of Article 13 of the RR(FS)O, which deals with fire-fighting and fire detection, can be summarised as follows:

☑ where necessary in order to safeguard the safety of relevant persons, the responsible person must ensure that the premises are equipped with appropriate fire-fighting equipment and with fire detectors and alarms; and any non-automatic fire-fighting equipment is easily accessible, simple to use and indicated by signs

☑ the responsible person must take measures for fire-fighting in the premises, suitable to the nature of the activities carried out there and the size of the undertaking, and of the premises concerned, and in doing so:

– nominate competent persons to implement those measures, and ensure that the number of such persons, their training and the equipment available to them are adequate

– arrange any necessary contacts with external emergency services, particularly as regards fire-fighting, rescue work, first aid and emergency medical care.

The enforcing body for fire safety on construction sites will often be the Health and Safety Executive (HSE), or the Local Authority (LA) on smaller sites. However, as outlined in the chart on the next page, there will be situations where it will be the local FRS.

C15

Fire prevention and control

For the purpose of demonstrating which body has the responsibility for enforcement, sites are divided into three types.

Type 1	Type 2	Type 3
Stand-alone sites, with a continuous perimeter, upon which the only activity taking place is construction. The HSE or LA will be the enforcing body for: - the management of GFP - process fire risks.	Shared occupancy sites* or premises with a breached separation (not fire or smoke proof) between the construction site and other activities. - The HSE or LA will be the enforcing body for the management of GFP. - The FRS will be the enforcing body for the management of process fire risks.	Shared occupancy sites* or premises with an unbreached, fire-resisting separation between the construction site and other activities. The HSE or LA will be the enforcing body for: - the management of GFP - process fire risks.
All types The FRS will be the enforcing body for fire safety with regard to temporary accommodation units (for example, site cabins) and other premises (such as storage areas and company offices) that are outside the site perimeter.		

It is anticipated that the vast majority of sites will be either Type 1 or 2. For a site to be categorised as Type 3, any fire occurring on it (or in the other occupied parts of the premises) must have absolutely no effect on the occupants of the other parts of the premises. Therefore, a site could be categorised as Type 2 if:

- ☑ there is a fire door in a fire-proof dividing wall between the two parts of the premises, because there is the potential for pedestrian traffic in either direction in the event of a fire on either side of the wall
- ☑ flames and smoke from a lower floor could enter the floors above, causing an evacuation, with either floor being the construction site.

* Examples of a shared occupancy site are where:

- ☑ one floor of a multistorey office block is a construction site but other floors remain occupied
- ☑ some houses are already occupied on an uncompleted housing estate
- ☑ an extension is being built on a domestic property and the owner remains in residence
- ☑ part of the sales floor of a department store is being refurbished but the public and store workers have access to the parts of the floor outside the hoarding.

With regard to fire safety on construction sites, readers should refer to:

- ☑ Chapter 2 for the requirements of the Construction (Design and Management) Regulations, with regard to fire safety
- ☑ Chapter 4 for the principles of risk assessment, which includes assessing the risk of fires.

Fire prevention and control

With regard to other premises, the Regulatory Reform (Fire Safety) Order requires a responsible person to be appointed by the employer to:

- ✓ take such general fire precautions as will ensure, so far as is reasonably practicable, the safety of any of their employees
- ✓ in relation to relevant persons who are not their employees, take such general fire precautions as may reasonably be required to ensure that the premises are safe.

A relevant person is any person who is or may be lawfully on the premises and any person in the immediate vicinity who is at risk from a fire on site or in site offices.

The term *general fire precautions* covers:

- ✓ reducing the risk of fire and spread of fire
- ✓ establishing methods of escape
- ✓ measures for securing the means of escape so that they can be safely and effectively used at all times
- ✓ measures in relation to the means of fighting fires on the premises
- ✓ measures in relation to the means for detecting fires and giving warning in case of fire
- ✓ arrangements for action to be taken in the event of a fire, including measures relating to the instruction and training of employees, and measures to mitigate the effects of a fire.

The Construction (Design and Management) Regulations mirror the requirements of the Regulatory Reform (Fire Safety) Order under the headings of:

- ✓ prevention of risk from fire
- ✓ emergency procedures

- ✓ emergency routes and exits
- ✓ fire detection and fire-fighting.

Contractors are required to pass on details of (fire) risk assessments to the principal contractor if there is a significant risk of fire arising out of the contractor's work.

Fire prevention

The precautions necessary to prevent the outbreak of fire must be actively managed; it cannot be left to chance.

A fire safety plan should be drawn up by someone competent to do so and it should be updated as the project progresses (for example, if new hot-works activities start or escape routes are changed). On larger sites, a competent member of the management team should be nominated as the fire safety co-ordinator.

A build up of combustible material posing a fire risk

Fire prevention and control

There is a very real risk of fire at all stages of most construction projects since:

- ☑ many activities involve the use of flammable materials whether they are solid (such as timber), liquid (such as vehicle fuels) or gaseous (such as LPG)
- ☑ some activities require or produce a source of heat, including equipment that has a naked flame.

Conditions required for a fire to start

Fire requires three elements to make it burn: fuel, heat and air. Without any one of these three, combustion cannot be sustained. However:

- ☑ flammable materials (fuel) will be present on site for much of the time
- ☑ on most projects there will be sources of heat at some time, even if only an office heater or using a grinder
- ☑ there is nothing that can be done about excluding the air around us.

The fire triangle

Given that it may well be impractical to exclude any of the above elements of a fire, the practical method of preventing fires is to ensure that flammable materials and sources of heat are kept well apart.

Hot-work permits

When activities are being carried out that create an enhanced risk of fire, the method by which the work is carried out should be controlled by a permit to work, usually in the form of a hot-work permit.

The conditions of such a permit will specify the fire prevention measures that must be taken before, during and after the work, such as:

- ☑ the need for a serviceable fire extinguisher of an appropriate type, located at the place where the work is being carried out
- ☑ a requirement that the person carrying out the work has been trained in the selection and use of fire extinguishers (*see also Staff training on page 217*)
- ☑ the need for fire checks during the work and for a period of at least one hour (or more depending upon risk assessment)
- ☑ the need for fireproof screens or mats to protect adjacent flammable materials or people passing by.

Managing fire risks

Identifying fire risks

Many fires on site are caused by tools and equipment that produce a naked flame, sparks or hot metal (such as blowlamps, oxyacetylene or oxypropane torches and welding); this is true irrespective of whether the work on site is new work, maintenance, repair or demolition.

Fire prevention and control

Hot works are carried out, for example, by on-site fabricators and steel erectors, plumbers, roofers and painters, sometimes in confined spaces and often near to flammable materials.

Many substances used on site (such as LPG, solvents or paint) are either flammable or give off a flammable vapour.

Protective coverings can contribute to the overall fire load – install vulnerable features as late as possible and ensure coverings are to flame-retardant specifications wherever reasonably possible.

High wattage, halogen flood lamps, which can produce sufficient heat to ignite dry materials, have been the cause of many fires. Where possible, alternative methods of lighting should be used, otherwise consider only using under hot-work permit conditions.

High risk timber frame buildings require additional fire prevention measures

Site fire safety plan

A fire safety plan must be drawn up to outline the measures that must be taken to:

- reduce the chances of a fire breaking out to the lowest practical level
- ensure the safe escape of everyone on site if a fire does occur.

Included within the plan should be:

- the identity of staff with responsibility for fire safety
- locations and type of fire-fighting appliances and fire alarm bells
- locations of assembly points
- fire escape routes
- arrangements for the storage of highly flammable material
- details of hot-works permit scheme.

Fire prevention measures include:

- carrying out a fire risk assessment for every job where there is a significant risk of fire
- implementing control measures to eliminate the risk of fire or reduce it to an acceptable level
- continually monitoring risk to ensure control measures remain effective as the project progresses
- clearly labelling fire points and position in locations identified during the fire risk assessment
- training staff in fire awareness (and possibly fire-fighting)

Fire prevention and control

A good fire point on site

- ✓ wherever possible, using alternative, non-flammable or less flammable materials for the job (the Building Regulations, designers and specifiers have a part to play in this)
- ✓ keeping only enough flammable material at the place of work for immediate needs
- ✓ storing flammable materials well away from sources of heat
- ✓ storing highly flammable substances in compounds, as required by the relevant legislation

Poorly stored fire extinguishers are no good if there is a fire

- ✓ not allowing flammable waste material to accumulate and disposing of it safely
- ✓ keeping flammable materials (solids, liquids or gases) away from sources of heat, for example:
  – welding and other hot work
  – heaters in site cabins and so on
  – electrical or other sparks
- ✓ keeping ventilation equipment clean, unobstructed and properly maintained
- ✓ keeping fire-resistant doors closed where installed
- ✓ not leaving equipment which incorporates an exposed flame (such as bitumen boilers), unattended at any time whilst they are alight.

Emergency procedures

In the event of a fire occurring, it is essential that the alarm is raised as quickly as possible so that everyone is aware and can quickly and safely reach a place of safety.

To achieve this:

- ✓ a means of fire warning must be provided on site and in site offices (hand bells, klaxons and manually or electrically operated sounders may be suitable if they are clearly audible above background noise in all areas and can be readily identified as being a fire alarm)
- ✓ written emergency procedures must be displayed in prominent locations within offices and should include:
    - instructions for raising the alarm
    - instructions to report to the nearest assembly point
    - information as to the whereabouts of the assembly point
    - the locations of fire escape routes.

In case of fire – no matter how small – or if a fire is suspected:

- ✓ raise the alarm
- ✓ call the emergency services
- ✓ close doors and windows, if possible, to prevent the spread of fire
- ✓ evacuate the site
- ✓ make sure that everyone is accounted for.

Anyone can call the emergency services when the alarm is heard, giving the address of the site and any directions that are necessary. It is better that the emergency services are informed of the fire by several people, rather than not at all.

Means of escape

Adequate means of escape must be provided to enable all employees and visitors to reach a place of safety should a fire occur.

Fire escape routes can be blocked unintentionally

Fire prevention and control

When considering the means of escape from a building, an employer should consider the following points:

- as part of a fire safety plan, dedicated escape routes should be identified, clearly signed and adequately lit
- all emergency exit/directional signs should be clearly visible and kept unobstructed and should conform to the Health and Safety (Safety Signs and Signals) Regulations and BS 5499
- signs should be positioned where the escape route changes direction or level (the signs must indicate the shortest safe route to a place of safety)
- signage and routes must be regularly reviewed to ensure they remain effective.

Burning waste

Bonfires should not normally be allowed on site. There should be alternative arrangements for the proper disposal of rubbish and waste. Environmental factors and/or the Local Authority may make the controlled burning of rubbish on site an unacceptable practice.

Generally bonfires are only acceptable in limited situations (such as the burning of vegetation as part of site clearance for road building projects).

Smoking restrictions

The Smoke-free (Premises and Enforcement) Regulations have introduced a legal ban on smoking in 'enclosed' and 'substantially enclosed' workplaces in England, with similar legislation covering Wales, Scotland and Northern Ireland.

This includes all forms of normal site accommodation (such as offices, canteens and toilet units) and other areas of the site that are categorised as 'enclosed' or 'substantially enclosed', as defined in the legislation.

An *enclosed* workplace is one that has a roof or ceiling and (except for passageways, doors and windows) is wholly enclosed, whether on a temporary or permanent basis.

A *substantially enclosed* workplace is one that has a roof or ceiling but there are permanent openings in the walls, the combined area of which is less than 50% of the total wall area. When calculating the total open area of any workplace, doors and/or windows that can be closed must not be counted.

Smoking in enclosed and substantially enclosed workplaces is now banned by virtue of different legislation in England and Wales, Scotland and Northern Ireland. Broadly, the differing legislation has similar aims in that:

- the act of smoking in smoke-free premises is an offence
- the act of permitting smoking in smoke-free premises (or a smoke-free vehicle) is an offence
- generally, vehicles that are used by more than one person must be smoke-free
- smoke-free premises must have an official 'No smoking' sign with the approved text, prominently displayed at each entrance
- smoke-free vehicles must have at least one official 'No smoking' sign prominently displayed.

Since smoking in the open air is allowed by law, provision for the safe disposal of smoking materials still has to be made. Carelessly discarded cigarette ends and matches have the potential to cause fires on site and, therefore, smoking in the open air should only be allowed in areas where it is acceptable from a fire-safety point of view.

The official 'No smoking' notice must be clearly displayed in any area where smoking is not allowed, including all entrances to all enclosed working places.

Fire prevention and control

Furthermore, the ways in which the smoking ban affects any particular site may vary as construction progresses. For example, whilst the ban will be total in site offices, canteens and other welfare facilities, it might be allowed outdoors (for example, where steelwork is being erected). However, once the steelwork is clad, the area inside the structure becomes enclosed and smoking must be prohibited. It will therefore be necessary to monitor the changes in the fire-risk areas and review the areas where smoking is allowed accordingly.

Site rules should ensure that smoking is prohibited for an appropriate period at the end of each working day (for example, the last hour). This will allow any developing fire to be discovered and dealt with before the site closes at the end of the day.

Trespassers

Children and other trespassers may start fires on site. Sites should, as far as possible, be secured against trespassers. In every case, combustible materials should be cleared away on a regular basis (daily) and not left lying around. Storage areas for flammable liquids and gases must be secured during non-working hours.

Everyone should know the correct action to take if they discover a trespasser on site during working hours.

Site offices and other accommodation

The risk of fire can arise from the use of heating and cooking appliances if they are not sited and installed correctly, adequately maintained or are not suitable for the intended use or location. Fuel supplies for gas-fired appliances, especially propane or butane, should be kept secured outside the building and piped in through fixed pipework. Any flexible pipework should be kept as short as possible, and used only for the final connections.

Combustible material should be kept well away from heaters and stoves. The practice of drying wet clothing in front of heaters must be prohibited. Care must be taken to see that newspapers, clothing or other combustible materials are not allowed to build up around such heaters.

All heaters, cookers and any other gas-powered appliances must be turned off at the end of each working day. Portable electrical apparatus should be switched off, unplugged and disconnected from the mains supply.

Fire action notices and fire escape signs should be installed as necessary.

Highly flammable liquids and liquefied petroleum gases

Highly flammable liquids (HFLs) and liquefied petroleum gas (LPG) are used extensively in the construction industry.

Spillage of HFLs or leakage of LPG will create a significant risk of fire if either come into contact with a source of ignition. To maintain fire safety, it is essential that such liquids and gases are:

- ☑ used safely
- ☑ handled and transported safely
- ☑ stored safely.

All three situations require that suitable fire extinguishers are located nearby.

Fire prevention and control

Highly flammable liquids

Any liquid that gives off a vapour that can be ignited at a temperature below 32°C is classed as a highly flammable liquid. Common examples include:

- petrol
- thinners
- solvents
- adhesives.

Some precautions to be observed when using HFLs are:

- only get out the quantity that is likely to be used in one day, or less, if practical and keep the remainder in a fireproof store
- keep containers closed at all times
- dispense or decant HFLs in designated areas and over drip trays to avoid spillage into metal storage bins
- quickly clean up any spillages that occur and safely dispose of the material used to soak up the spillage
- provide suitable fire-fighting measures in case of emergency
- prohibit anything that could create naked flames, sparks or other means of ignition
- treat empty drums and containers with care (they will contain a mixture of vapour and air that could be an explosion hazard)
- HFLs or solvents should not be used to clean floors or surfaces because this could cause a flammable atmosphere – always use a detergent cleaner instead
- when using a HFL or solvent to clean or soak brushes, always use a container with a lid and keep it closed or carry out the cleaning in the open air.

Do not bring petrol into timber frames or enclosed buildings – use designated refueling points or better still another type of powered equipment

Liquefied petroleum gases

Liquefied petroleum gas is stored in cylinders at high pressure, which keeps it in a liquid state. When released to the air, the pressure is reduced and it reverts to a gas. Because it is stored as a liquid it is essential that LPG cylinders are secured in an upright position when being used, transported or stored.

LPG is used extensively in the construction industry as a means of producing heat during some work processes (such as heating bitumen boilers, soldering, or as a means of heating or cooking).

There are two types of LPG, propane and butane, both of which operate at different pressures. It is, therefore, essential that no attempt is made to use the equipment designed for use with propane on butane or vice versa.

Another hazardous feature of LPG is that it is heavier than air. If allowed to leak, the gas will sink to the ground and find the lowest point (such as drains, excavations, cellars and so on). It will form an explosive reservoir just waiting for a source of ignition.

LPG and air in the correct concentration is an explosive mixture. One litre of liquid LPG can produce more than 250 litres of gas. If that quantity were to be ignited, the resultant explosion could destroy a building. If LPG cylinders are being engulfed by a fire there is a serious risk of the cylinders exploding. The area should be evacuated and a cordon formed to keep people out of the area.

Leaks of LPG can be detected by the smell of the gas or by frosting around the leak.

Frosting on the cylinder may indicate that the gas discharge rate is too high and there is a need to reduce the gas flow or couple-up another cylinder by means of a manifold. Where there is doubt supervisors should seek advice.

Never use a naked flame to detect a leak.

If the leak can be isolated or turned off, do so. However, if that is not possible, and it is safe to do so, the cylinder should be moved into an open, evacuated space and the fire service called. If the cylinder is involved in a fire, evacuate the area immediately and inform the fire service of the situation.

When not in use, the cylinder valve must be closed to prevent the possibility of leakage. Full cylinders must be kept in a store that will allow any leakage to disperse.

Fire-fighting

Knowing what to do in the event of a fire is essential. Using the wrong fire-fighting equipment can turn an already serious situation into a deadly one. It is crucial that only people who have been trained to select and use fire extinguishers attempt to fight fires, and they can recognise when the situation is sufficiently serious that fire-fighting must be left to the Fire and Rescue Service.

Staff training

The following points should be emphasised to all staff with regard to fire extinguishers and fire-fighting training if fighting of small fires is to be allowed.

The decision to allow the safe fighting of fires should be taken via a risk assessment, taking into account requirements of Regulation 41(5) of CDM:

Every person at work on a construction site shall, so far as is reasonably practicable, be instructed in the correct use of any fire-fighting equipment that it may be necessary for them to use.

The Fire Safety (Employees' Capabilities) (England) Regulations require that employers in England ensure that all tasks related to fire safety are only allocated to employees with the necessary skill, knowledge and experience to carry them out safely. This will inevitably require that appropriate training is provided for those engaged in hot works, fire wardens, site managers and so on.

If the fighting of small fires is allowed by staff who have been trained to use fire extinguishers, the following points must be emphasised.

Fire prevention and control

10 rules for fighting fires

1. Do not put yourself in danger.
2. If you discover a fire, raise the alarm first and ensure that an evacuation is underway before fighting the fire.
3. Only use an extinguisher if it is safe and you have been trained to do so.
4. Do not let the fire come between you and your escape route. You may become trapped if the fire develops.
5. If the extinguisher does not appear to be working or is ineffective on the fire get out immediately.
6. If the fire starts to develop or gets out of control get out immediately.
7. Do not misuse fire extinguishers (for example, in boisterous play or dampening down).
8. Do not move fire extinguishers from their allocated positions.
9. Do not use fire extinguishers as door stops.
10. Immediately report (or replace) any fire extinguishers that appear to have been used, misused or damaged.

To comply with regulation 41(5) of CDM, based upon a risk assessment, each company must decide who, from amongst its staff, will be trained to use fire extinguishers, and provide guidance upon the point at which fire-fighting must be abandoned and left to the Fire and Rescue Service.

Portable fire extinguishers

Where there is a realistic possibility that staff may have to use a fire extinguisher, they should be trained in their use. In selecting staff for training, thought must be given to the size and weight of extinguishers.

There are several different types of fire extinguisher, each indicated by a different colour bank or panel on the extinguisher body. The colours indicate the different substances that they contain and therefore the different types of fire for which they are suitable.

With older fire extinguishers, the contrasting colour covers the whole of the body. Both old and new type water-filled extinguishers are completely red.

The chart on page 220 shows which types of fire extinguisher should be used on different types of fire and, equally importantly, which fires some extinguishers must not be used on.

All fire-fighting equipment must be maintained and inspected regularly, and all such inspections recorded in the appropriate register.

Fire blankets

Fire blankets are normally found in catering facilities. They are usually sufficient to deal with small, contained fires involving frying pans or other cooking vessels.

Anyone who may have to use a fire blanket should be trained to do so, because it involves placing the blanket gently over the burning pan of oil, which requires getting close to the fire and behaving in a controlled manner. Before a fire blanket is used, the gas or electricity supply should be turned off where circumstances permit.

Fire prevention and control

To use a fire blanket:

- ✓ pull the blanket from its container and wrap the corners of the blanket around your hands, making sure that the hands and forearms are completely covered

- ✓ hold the blanket at chest level and gently place it over the burning container to exclude the air from the fire (do not throw the blanket as it may miss the burning container or cause it to spill)

- ✓ leave the blanket in place until the container has cooled down (do not lift one corner to check if the fire is out as this may allow enough air in to reignite the fire).

A fire blanket can also be used to put out clothing that is on fire.

Fire hydrants

Adequate water for fire-fighting must be available. This should be achieved by utilising the fire hydrants fixed to existing street mains or by providing a separate supply. The amount of water likely to be required should be discussed with the FRS as part of the liaison process.

All fire hydrants must be clear of obstructions and suitably marked. Particular care should be taken to ensure that site plant, delivery lorries or workers' cars are not parked close to or over hydrants.

It should be noted that it is an offence for any vehicle to obstruct a fire hydrant and that the FRS has the power to initiate the prosecution of offenders.

Fire wardens

Where the complexity of the site and/or the accommodation units make it difficult to establish that everyone has been evacuated from the site, it will be necessary to appoint fire wardens. Each fire warden should be allocated an area to 'sweep' in the event of an evacuation to make sure that no-one remains on site. Fire wardens can also be used to monitor the effectiveness of the general fire precautions and contribute to the ongoing development of the fire safety plan.

Fire prevention and control

Types of portable fire extinguisher and what to use them on

Make yourself aware of the instructions on the fire extinguisher before using it

Fire class	Substances, materials, and so on	Water (red label)	Foam (cream label)	Carbon dioxide (CO_2) (black label)	Dry powder (blue label)
A	Carbonaceous and organic materials, wood, paper, rag, textile, cardboard, common plastics, laminates, foam.	YES, excellent	YES	Difficult to use outdoors in windy conditions. For small fires only if no water available	YES
B	Flammable liquids, petrol, oil, fats, adhesives, paint, varnish.	NO	YES, if liquid is not flowing	YES, but not ideal	YES
C	Flammable gas: LPG, butane, propane, methane, acetylene.	NO, not effective on gas flame but will cool the area and put out secondary fires	YES, if in liquid form. (Seek specialist advice.)	NO	YES, excellent
D	Metal, molten metal, reactive metal powder.	NO	NO	NO	YES, trained person – if no explosive risk. Special powders are available, but DRY sand or earth may be used
Electrical	Electrical installations, typewriters, VDUs, computers, photocopiers, televisions, and so on.	NO	NO	YES	YES, but not ideal. Or switch off electricity and deal with as an ordinary fire
F	Fires with cooking appliances that involve vegetable or animal fat.	NO	YES, with F rating only	YES, with F rating only	YES, with F rating only

Note: dry powder may not penetrate spaces or behind equipment; light water foam (AFFF) may be used instead of water or foam. Extinguishers used to control Class B fires will not work on Class F fires because of the high temperatures generated.

So, in summary, what needs to be in place if a fire occurs?

- ✓ An effective and secure means of escape.
- ✓ A process to detect fires and give warning.
- ✓ The appropriate equipment to fight fires.
- ✓ A process and equipment to mitigate the effects of fire.
- ✓ A process and equipment to reduce the risk of fire and its spread.
- ✓ Employees who have been instructed and trained in the action to take.

For further information refer to GE 700 *Construction site safety*, Chapter A03 Construction (Design and Management) Regulations and Chapter C02 Fire prevention and control.

Fire prevention and control

16

Electrical safety

What your employer should do for you	224
What you should do as a supervisor	225
Relevant legislation	226
Electricity at Work Regulations	226
Managing the risks	227
Safe system of work	227
Low voltage hand tools	228
Mains-powered hand tools and equipment	228
Safe working close to overhead cables	230
Safe working close to underground cables	231
Site lighting	231
Dealing with electric shock	232

Electrical safety

What your employer should do for you

1. Ensure that the electricity supply is tested and inspected regularly in line with regulations.
2. Arrange for electrical tools and equipment to have regular inspections and PAT testing, and that records are kept.
3. Plan that the preference for using 110 V or battery-powered tools is followed by the company.
4. Put controls in place to prevent unauthorised use, alteration or repair to the electrical supply, tools and equipment.
5. Ensure that isolation points are clearly identified.
6. Provide and maintain adequate site and task lighting.
7. Ensure that in the event of power failure there is a back-up plan in place to ensure safe egress.
8. Provide the work team with training in actions to be taken in the event of electrocution.
9. Ensure that the length of trailing cables and pressure hoses is minimised to prevent danger.
10. Put arrangements in place to check for overhead and underground power cables and provide appropriate measures.
11. Implement permits to work and isolation systems for work to electrical supplies, and ensure that they are monitored regularly.

Electrical safety

What you should do as a supervisor

Checklist	Yes	No	N/A
1. Test and inspect the electricity supply regularly in line with regulations.			
2. Present electrical tools and equipment for regular inspections and PAT testing, and keep records up to date.			
3. Use 110 V or battery-powered tools as a preference.			
4. Prevent unauthorised use, alteration or repair to the electrical supply, tools and equipment.			
5. Check that isolation points are clearly identified and unobstructed.			
6. Maintain site and task lighting.			
7. Check that the back-up plan to ensure safe egress in the event of power failure is working.			
8. Check that the work team have training in actions to be taken in the event of electrocution.			
9. Minimise the length of trailing cables and pressure hoses to minimise danger.			
10. Mark up and maintain warnings for overhead and underground power cables.			
11. Issue permits to work and isolations systems for work to electrical supplies, and ensure that they are used correctly.			

C 16

Electrical safety

Relevant legislation

Management of Health and Safety at Work Regulations

These regulations require that employers carry out a suitable and sufficient risk assessment of the work they do and put in place measures to control the risks arising from the work. In the context of this chapter, this will mean that procedures and practices are put in place to eliminate or reduce the risk of anyone being harmed as a result of coming into contact with an electrical supply.

For further information on carrying out risk assessments refer to Chapter A04 Risk assessments and method statements.

Provision and Use of Work Equipment Regulations (PUWER)

Electrical equipment and hand tools are classified as *work equipment* as defined under PUWER. As such, employers have a duty to ensure that electrical hand tools and equipment:

- ☑ are suitable for their intended purpose
- ☑ are well-maintained and inspected as necessary
- ☑ are fitted with suitable controls
- ☑ are fitted with guards as necessary
- ☑ can be effectively isolated from the supply
- ☑ carry appropriate markings and warning notices.

Furthermore, employers have a duty to ensure that users of work equipment receive adequate health and safety information, including written instructions, where appropriate, on the use of the equipment.

For further information on PUWER refer to Chapter C17 Work equipment and hand-held tools.

Electricity at Work Regulations

The requirements of these regulations cover fixed electrical distribution systems rather than electrical hand tools and equipment. The supply system that feeds mains power onto the site and distributes it to site offices, canteens and so on, must be installed in accordance with these regulations.

Electricity has a hazard warning sign like this

Each year, construction industry workers are killed or injured as a result of unsafe electrical installations, unsafe practices or defective tools. The danger arises as follows.

- ☑ Electricity is invisible; it never tells you that it is coming and it can kill you.

Electrical safety

- Low levels of current (just 1 mA) can be enough to throw you off balance and make you fall. Higher levels of current cause muscles to spasm and can make it impossible to let go of the object you are holding. With 50 mA, the skin will burn at the point of contact, the heart is affected and death can occur.

- Electricity can cause sparks. In places where there is airborne dust or flammable vapours, those sparks can cause explosions.

- Site electrical distribution systems are temporary, often operate in harsh (including wet) conditions and are often modified as the demand for supplies change.

- Overhead cables are often un-insulated, meaning that the electricity will flow through any metal object that comes into contact with them (such as a raised excavator bucket).

- Overhead cables sometimes carry high voltage supplies, which means that the electricity can jump (arc) to anything nearby that will conduct electricity without it actually touching the cable. The more moisture there is in the air, the further the electricity can jump.

- The existence of underground cables is sometimes not established, or they are not accurately located before excavation of the ground starts.

A little knowledge is often sufficient to make electrical equipment function but a much higher level of knowledge and experience is usually needed to ensure safety.

Managing the risks

Treat electricity with respect. Always assume that an electrical supply is live unless you have proof that it is not.

Do not tamper with the site electrical distribution system unless you are competent to do so, it is your job and you have been authorised to make the necessary changes.

Contractors of the electricity supply company must install the main supply from off site to the site distribution board. Further distribution to site offices and other accommodation must be installed by competent electrical contractors. All mains (230 V) circuits must be supplied through residual current circuit breakers located in the fuse board.

The factors that will reduce the risk of injuries from the use or proximity of electricity are discussed below.

Safe system of work

A risk assessment will identify any risk of electric shock and identify the measures that must be taken to ensure that work can be carried out safely. In some circumstances a permit to work system will be necessary to ensure that the electrical supply is locked-off before work starts and remains locked-off until the work is complete and it is safe to restore power.

Wherever practical, work should not be carried out on live equipment or near to a live electrical supply. If possible the power should be switched off. The Electricity at Work Regulations require that work on or near exposed live electrical conductors can only be carried out when:

- it is unreasonable for the supply to be made dead

- it is reasonable for live working to be carried out

- suitable precautions are taken (for example, RCDs, PPE, screening and so on).

Electrical safety

Low voltage hand tools

The safest electrical hand tools operate from low-voltage batteries. An increasing range of battery-powered hand tools are becoming available.

All non-battery electrical hand tools used on building or construction sites should operate from a 110 V supply. The way that the system is wired means that the supply voltage is effectively 55 V, which reduces to a safe level the severity of any shock that occurs.

The 110 V supply will usually be derived from a transformer that is plugged into a 230 V (mains) supply or from a portable generator. The power lead, casing or plug will be yellow in colour signifying that it is 110 V equipment. A round, **blue** plastic plug or a domestic 3 pin plug signifies that the equipment is 230 V (mains) powered.

415 V (three-phase) cable casing and plugs and sockets should be **red**.

Plugs and sockets should be manufactured to BS EN 60309-2:1999. This standard will prevent accidental or intentional connection of tools to power supplies of the wrong voltage.

Mains-powered hand tools and equipment

Wherever possible, the use of electrical hand tools that operate directly off the mains supply must be avoided and battery or 110 V tools used instead.

However, where the use of a mains-operated tool or other equipment is unavoidable, possibly because there is not a 110 V version, a portable residual current device (RCD) must also be used. RCDs are often known as power breakers; they plug into the mains socket and have an inbuilt socket into which the tool is then plugged.

An RCD will detect an electrical fault and disconnect the supply to the tool very quickly and before an electric shock would be sensed by the user.

> RCDs have a test button that should be operated at the start of each day to check the correct operation of the device.

Proprietary battery banks can be hired to facilitate safe and secure recharging

Electrical safety

The supplies for other mains-operated electrical equipment (such as office and catering equipment), should be fed through an RCD in the main fuse board. The operation of these should be checked weekly by operating the test button.

Care of electrical hand tools and other equipment

Working with damaged electrical tools or equipment can spell danger. Before starting work, a brief visual inspection should always be made for:

- ✓ cables that are cut, abraded or pulled out from the plug
- ✓ damaged plugs
- ✓ damaged casings.

Wires pulled from a plug can be caused by:

- ✓ pulling or dragging cables across the ground
- ✓ picking the tool up by its cable
- ✓ overstretching the cable
- ✓ flex clamps in the plug not secured tightly enough.

If you find a fault, report it and ensure that the equipment is repaired or replaced.

Electrical hand tools should be subjected to a periodic test of their electrical safety, known as **portable appliance testing (PAT).** The PUWER Regulations do not specify what testing needs to be done, by whom or how frequently (for example, they do not make it a legal requirement to test all portable electrical appliances every year). This allows the duty holder to select precautions appropriate to the risk. A sticker should be affixed to hand tools that pass the test, showing when the next test is due. Electrical hand tools used by construction workers can be roughly treated and it is recommended that they should have a combined inspection and test before they are first used and then every:

- ✓ three months (110 V tools)
- ✓ one month (230 V tools)
- ✓ one month (portable RCDs)
- ✓ three months (fixed RCDs)
- ✓ year (equipment in site offices).

These suggested frequencies for inspection and testing are not legal requirements. For further information see the HSE publication *Maintaining portable electrical equipment* (HSG107).

If an electrical hand tool becomes defective whilst it is in use, the user should stop work immediately and disconnect the tool from the supply. It should not be used again until it has been repaired.

In many cases a user will be capable of replacing a blown fuse in the plug of a 230 V tool, but other repairs must be carried out by someone who is competent to do so. If a fuse does blow:

- ✓ unplug the tool from the supply
- ✓ look for obvious signs of damage
- ✓ replace the fuse with another of the correct rating if there are no obvious signs of damage
- ✓ accept that the tool is defective if the second fuse blows
- ✓ if necessary, label the tool as defective and put it where no-one else will be tempted to try and use it.

Electrical safety

Safe working close to overhead cables

Electricity carried by overhead cables, which are generally uninsulated:

- ✓ will flow through any metal object that comes into contact with them (such as a metal ladder, a scaffold pole or a raised excavator bucket)
- ✓ may jump through the air (arc) to anything nearby that will conduct electricity.

Where overhead cables are within premises it may be possible to arrange with the owner or occupier to temporarily switch off the power. If the cables are part of the national power distribution network, isolation of the supply may not be possible or there may be strict time limitations.

If it is not possible for the power to be switched off, the power supply company must be consulted to establish the minimum safe working distance, from which a safe system of work can be developed. This will often involve erecting barriers and possibly introducing other control measures to ensure that nothing and no-one encroaches into the safe distance.

Minimum safe distances are usually accepted as:

- ✓ 9 m from cables carried on wooden poles
- ✓ 15 m from cables carried on metal pylons.

Source: HSE Publication HSG150.

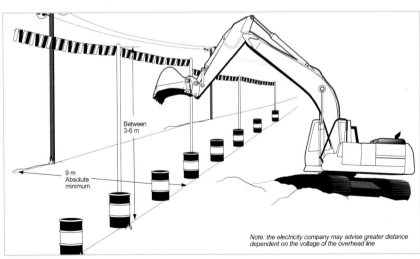

Note: the electricity company may advise greater distance dependent on the voltage of the overhead line

Electrical safety

Barriers are often formed using a line of concrete-filled drums with non-metallic warning posts, positioned parallel to the line of the cables. The minimum distance from the barrier to the nearest cable should be 6 m, although the electricity supply company may recommend a greater distance.

 Electricity can kill. The correct information, instruction, training and supervision can help to keep workers, and others coming into contact with them, alive.

Find out who owns the installation first and talk to them before you start work. (Is it possible to isolate or insulate?)

Safe working close to underground cables

Accidentally damaging underground cables has caused deaths and serious burns. Carry out these steps if the job involves penetrating the ground surface.

Six steps to safety
1. Obtain and understand current drawings.
2. Arrive, observe and plan.
3. Locate using a suitable CAT.
4. Mark up cable routes.
5. Hand dig trial holes.
6. Ensure compliance with HSG47.

 For further information on safe excavation practices refer to Chapter B08 Health and welfare.

Site lighting

The Provision and Use of Work Equipment Regulations (PUWER) require that:

 Every employer shall ensure that suitable and sufficient lighting is provided at any workplace.

The Construction (Design and Management) Regulations (CDM) require that adequate lighting is provided at every place of (construction) work and approach to the workplace, and that secondary (back-up) lighting is provided where failure of the primary lighting would result in risks to health or safety.

Therefore, these two legislative provisions place a duty on employers to ensure that well-planned and adequate lighting be installed on site where necessary for safe working and access.

To obtain such lighting conditions, you should consider:

- installing suitable lighting to provide the required level of illumination for the nature of the work being carried out
- mounting the lights at a suitable height above the work level to give the required spread
- positioning lights to avoid glare, dazzle and reflection, as far as is possible
- changing the position of lights as work proceeds

Electrical safety

- ✓ screening or shielding lights from reflective surfaces, on traffic routes and so on
- ✓ routing of lighting cables to avoid tripping or damage to the cable
- ✓ installing back-up lighting where appropriate.

Lights constitute a heat source and, therefore, a fire hazard. They also retain heat for a period after being switched off. Attempts must be made to avoid locating them in close proximity to combustible materials. Halogen floodlights get particularly hot and have been attributed as the cause of at least one major fire.

Temporary electrics fixed to soffit with power/light drops that also help minimise trip hazards

Dealing with electric shock

- ✓ Could you help someone who has had an electric shock? Would you know what to do?
- ✓ Do you know that what you do could make the difference between life and death?
- ✓ Do you know how to apply resuscitation correctly?
- ✓ Have your employees been trained to do it correctly?
- ✓ If you were to be the casualty, is there anybody who could do it for you?
- ✓ Do you have the 'Danger of electric shock' notices or posters available?
- ✓ Can these notices or posters be seen by everyone? Are they well positioned?
- ✓ Are they read and understood?

Electric shock action

If the casualty is in contact with what could be a live electrical supply, switch off power if possible and shout for help. Never assume the power has been turned off unless you have received categoric assurance that it has.

If the power cannot be switched off:

- ✓ do not touch the casualty with bare hands
- ✓ if practical and safe, move the casualty clear of the electrical source or the source away from the casualty with a broom or something else wooden – never use a metal object as this will conduct electricity through you
- ✓ seek prompt help from a first aider and get qualified medical assistance.

Electrical safety

 Never put yourself in danger.

 For further information refer to GE 700 *Construction site safety*, Chapter C03 Electrical safety.

Electrical safety

17

Work equipment and hand-held tools

What your employer should do for you	236
What you should do as a supervisor	237
Legislation	238
Construction plant and vehicles	239
Abrasive wheels	240
Cartridge and gas-fired operated tools	241
Gas-powered fixing tools	242
Woodworking machinery	243
Other tools and equipment	243

Work equipment and hand-held tools

What your employer should do for you

1. Arrange that all plant and equipment is checked and tested.
2. Ensure that statutory records are kept up to date and maintain registers and ensure that all work equipment is identified and marked.
3. Implement procedures for the inspection and repairing or replacing of equipment and tools and rectify any faults developed through misuse or neglect.
4. Ensure that drivers and operators are trained on specific plant or equipment and are aware of hazards associated with its operation.
5. Maintain up to date lists of authorised drivers and operators.
6. Provide relevant information in relation to specific items of plant where restrictions of use apply, where it is kept and by whom.
7. Monitor that plant is being used safely.
8. Ensure that traffic routes are observed by site vehicles as applicable.
9. Prevent any instance of overloading or over-stressing of plant that comes to light.
10. Ensure that roll-over protective structures (ROPS), falling object protective structures (FOPS) and driver restraint systems are fitted and used, if appropriate.
11. Monitor that operators of mobile or self-propelled plant have adequate visibility in all directions and all visibility aids are in good order.

Work equipment and hand-held tools

What you should do as a supervisor

Checklist	Yes	No	N/A
1. Inspect all plant and equipment and that it is tested.			
2. Maintain statutory records, keep registers updated and ensure that all work equipment is identified and marked.			
3. Arrange for the repairing or replacing of equipment and tools and rectify any faults developed through misuse or neglect.			
4. Check that drivers and operators are trained on specific plant or equipment and are aware of hazards associated with its operation and report shortfalls.			
5. Update lists of authorised drivers and operators.			
6. Retain and provide relevant information in relation to specific items of plant where restrictions of use apply, where it is kept and by whom.			
7. Check that plant is being used safely and stop work where necessary.			
8. Ensure that traffic routes are observed and kept clear of obstructions by site vehicles as applicable.			
9. Manage the prevention of any instance of overloading or over-stressing of plant.			
10. Check that roll-over protective structures (ROPS), falling object protective structures (FOPS) and driver restraint systems are used properly.			
11. Ensure that operators of mobile or self-propelled plant have adequate visibility in all directions and maintain segregation of plant and people.			

C 17

Work equipment and hand-held tools

Legislation

Provision and Use of Work Equipment Regulations

Under the Provision and Use of Work Equipment Regulations (PUWER), *work equipment* is defined as all equipment used in the course of carrying out work. Included within this definition are:

- ✓ hand tools (such as hammers, trowels, handsaws and pipe-benders)
- ✓ small plant (such as cement mixers and portable generators)
- ✓ construction plant (such as excavators, dumpers and mobile compressors)
- ✓ lifting equipment (such as tower cranes and gin wheels).

PUWER applies to all work equipment and sets the standards for the provision and safe use of such equipment. The primary objective is to provide all workers with suitable and safe equipment and to ensure its proper use.

Full compliance with the regulations should ensure that work equipment does not give rise to any health and safety risks. There is a duty on employers to:

- ✓ ensure that equipment is constructed or adapted for the purpose for which it is used or provided
- ✓ consider working conditions and risks to health and safety to persons where the equipment is to be used
- ✓ ensure that work equipment is properly maintained and inspected, as necessary, by a competent person
- ✓ ensure that dangerous parts are guarded to the extent possible whilst still enabling the equipment to be used for its intended purpose
- ✓ provide adequate information, instruction and training, where necessary, for the safe use of the equipment
- ✓ ensure, where appropriate, that work equipment is fitted with effective and clearly marked controls and that it can easily be disconnected from the source of power.

These regulations specifically require that, where necessary, mobile plant is fitted with:

- ✓ roll-over protection for the operator where otherwise there would be a danger of injury if the machine were to roll over (a roll-over cage or frame)
- ✓ seat belts where, otherwise, there would be a danger of the operator being thrown from the machine, or being crushed by it, if it were to roll over
- ✓ falling object protection where the operator would otherwise be at risk of injury from falling objects or materials.

Plant must be well maintained, suitable for the job in hand and used or operated by a competent person

Work equipment and hand-held tools

Management of Health and Safety at Work Regulations

These regulations require that employers assess the risks to health and safety arising out of their work activities and put control measures in place to eliminate the risks or reduce them to a safe level. In the context of work equipment this process will often involve:

- ✓ selecting the most suitable piece of work equipment for doing a job
- ✓ ensuring that it has been properly maintained, which may involve formal inspection and certification, as required for lifting equipment
- ✓ ensuring that those who have to use or operate work equipment are trained and competent. The degree of formal training required will depend upon:
  - the complexity of the work equipment
  - the potential for harm if it is not used correctly.

Establishing a person's competency may involve checking that they hold valid competency cards (such as CPCS (plant operators)).

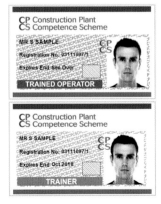
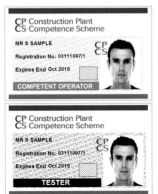

Associated legislation

Other chapters in this book provide details on the legislation that is specific to work equipment that is used for:

- ✓ working at height
- ✓ mechanical lifting operations (for example, cranes, slings and so on)
- ✓ personal protection (PPE and RPE).

Information, instruction, training and supervision

The provision of all necessary information, instruction, training and supervision must be readily available, written where appropriate, and easily understood. It must include:

- ✓ the conditions in which, and methods by which, equipment shall be used
- ✓ any foreseeable abnormal conditions and appropriate action
- ✓ any conclusions that can be drawn from experience with equipment
- ✓ safe working methods
- ✓ possible risks that may be found and subsequent precautions to be taken.

Construction plant and vehicles

Accidents involving plant and vehicles that result in serious injuries and fatalities are all too common. Many of these accidents are due to:

- ✓ the operator's limited range of vision when at the controls
- ✓ semi-automatic quick hitch releasing devices not having the retaining pin or bar replaced after attachment changes, or automatic devices not being checked for closure

Work equipment and hand-held tools

- ☑ passengers being carried on plant that is not designed to carry them
- ☑ untrained and/or unauthorised operators using mobile plant.

As a very minimum, all mobile plant should be equipped with:

- ☑ a reversing alarm
- ☑ at least one amber beacon that is switched on when the machine is in operation
- ☑ fully working lights and indicators
- ☑ extra driver vision aids (such as additional mirrors), to give 'all round' visibility or as near as possible to it
- ☑ roll-over protection, seat belts and falling object protection where appropriate.

Prevention and control measures

- ☑ Unauthorised use of machines, both during and out of site working hours, is easily prevented by implementing a system of close control for keys and other starting devices.
- ☑ Daily pre-user and weekly recorded checks of plant before start-up, checking that everything works as it should and levels and pressures (tyres) are correct, air storage tanks are drained and so on.
- ☑ Avoiding spills when re-fuelling and providing spill kits. If the plant is static, providing drip trays.
- ☑ Preventing overloading and unsafe loads.
- ☑ Stopping work if you suspect a problem.

For information on (logistics) plant movement, pedestrian segregation, traffic management and so on, refer to Chapter C18 Site transport safety.

Abrasive wheels

Diamond blades, angle grinders and disc cutting tools

These are very dangerous if not used correctly by trained and competent staff.

They can inflict a severe injury to the user, and others, in an accident. They operate at very high speeds and can injure by:

- ☑ the operator's hand coming into contact with the revolving wheel
- ☑ particles (such as hot metal or sparks) being thrown off while in use
- ☑ disintegration (shattering or loss of segments) of an overspeeding, damaged or incorrectly used disc, blade or wheel.

Injuries to hands through contact with a rotating wheel can be severe, although there is a limit to which the wheel may be guarded.

The most common causes of accidents involving abrasive wheels are:

- ☑ using the wrong type of disc, blade or wheel and incorrect mounting
- ☑ the fire hazard from the stream of hot sparks and the heat generated in the material being worked on
- ☑ overspeeding due to an incorrect disc, blade or wheel being used
- ☑ diamond blades losing segments, particularly if not used in line with manufacturers' instructions.

Work equipment and hand-held tools

Using abrasive wheels

Abrasive wheels should only be mounted by a trained and competent person who has been appointed by the employer and that appointment should be in writing.

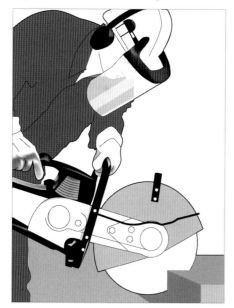

Wear the appropriate PPE and ensure the guards are properly set

Operators must:

- ☑ work on a firm, clean and unobstructed base
- ☑ maintain a firm grip when operating
- ☑ plan their own working posture and position to avoid injury in case of slippage or other unintended movement
- ☑ use adequate protection from flying debris for themselves and others in the vicinity
- ☑ ensure adequate support and stability of the material being worked on
- ☑ be aware of the health hazards (such as flying debris, dust, noise and vibration). This will usually involve:
  - keeping other people out of the area
  - wearing RPE, hearing protection, eye protection and gloves
  - not working in a confined area where the dust cannot disperse
  - using a dust collector or wet cutting that damps down the dust where these are a feature of the machine
  - ensuring that guards are in place, adjusted and secure
  - doing the job in short spells or by job rotation if hand-arm vibration is an issue.

Cartridge and gas-fired operated tools

Cartridge-operated tools

 These tools are potentially lethal in the hands of the untrained or foolish. They look and perform like guns.

They may be used for installing repetitive fixings on site and are either piston-operated by the cartridge (indirect) or with a cartridge operating the fixing device (direct). They may be high or low powered.

Work equipment and hand-held tools

The hazards most associated with cartridge tools are:

- ☑ lack of knowledge and training
- ☑ deliberate misuse, including boisterous play, by pointing the tool at someone
- ☑ poor maintenance.

There are two particular problem areas associated with the use of these tools – through penetration and ricochets.

Penetration is caused by:

- ☑ the cartridge being too powerful
- ☑ thin materials or voids in the structure being worked on
- ☑ changes in density of the material being worked on
- ☑ the material density not being ascertained before starting work.

Ricochets are caused by:

- ☑ the fixing passing straight through the material being secured
- ☑ second attempts at the same hole
- ☑ working on excessively hard materials (hardened steel or welded areas)
- ☑ not holding the tool square
- ☑ working too close to the edge of the material
- ☑ obstructions inside the material (reinforcing rods or dense aggregate).

Selection of the correct cartridge, trial fixings and checking behind the structure to be worked on will do much to prevent these hazards.

Careful alignment of tools, examination of structures and the use of low-powered, indirect-acting tools will also help to avoid these hazards.

Other safety aspects

Safety helmets, impact-resistant eye protection and ear protection should be worn at all times when cartridge-operated tools are being used.

Take care when using the tool since any recoil could lead to loss of balance. Always work from a scaffold or working platform if possible. Never use a cartridge tool while on an untied ladder.

In the event of a misfire, keep the tool pressed against the workface for at least 30 seconds then follow the manufacturer's instructions to the letter.

Operatives must be adequately trained in the use of equipment and made aware of the hazards. They should be tested for colour blindness, as cartridges are colour coded for identification.

Gas-powered fixing tools

Nail guns

Many of the principles for the safe use of cartridge-operated tools also apply to gas-powered fixing tools, which use a canister of pressurised gas (a fuel cell) as a propellant. Generally, gas-powered fixing tools are used for firing fixings into softer materials (such as timber). However, in untrained hands they can be as dangerous as cartridge-operated tools.

The implications of a misfire when using a gas-powered fixing tool are not as serious as when using a cartridge-operated tool and it is usually safe after a misfire to attempt to make the next fixing immediately. The battery and fuel cell must be removed prior to attempting to remove a blockage.

Work equipment and hand-held tools

 Only persons over 18 years of age should be allowed to use power-operated fixing tools. Only mature and responsible people should be selected for this work.

Woodworking machinery

Woodworking machines (such as portable bench circular saws, portable hand-held circular saws and planing machines) are all found on building and construction sites.

They are also the cause of many accidents. Poorly adjusted guards and the failure to use push-sticks contribute to many injured or lost fingers.

Only competent persons or trainees under direct supervision should operate any woodworking machine.

A sound, firm and level base with a clear and well-ventilated working area must be established prior to work starting.

All guards and safety devices should be fitted, correctly adjusted and fully operable all the time the machines are running. The safety distances of the guards from the blades should be closely monitored.

Machines must not be left running and unattended. The operator should not leave a machine until it has fully run down.

In December 2008 it became a requirement that all woodworking machines with a rundown time of 10 seconds or more must be fitted with a braking device, unless the effect of braking could be detrimental to the integrity of the machine.

Other tools and equipment

Petrol-driven hand tools

Petrol must be kept in small quantities in approved containers, refuelled only in well-ventilated areas using a funnel without the engine running or while parts are still hot. Exhaust fumes are toxic and must not accumulate in enclosed or confined spaces.

Electrical hand tools

All electrical hand tools used on site should only be 110 V or battery powered. Before every use you should carry out a brief visual inspection of the power lead and plug, casing, switches, triggers and guards. You must switch off and remove the plug before carrying out any adjustment.

Compressed air-powered tools

Compressed air tools are attached to a compressor using air hoses. Tools include heavy duty breakers, soil picks, concrete scabblers and pokers.

Always check hose fittings are tight and secure before use. High pressure air hoses can cause serious injury if they break away from the compressor or tool. Whip checks should be used on every hose joint to prevent this.

C 17

Work equipment and hand-held tools

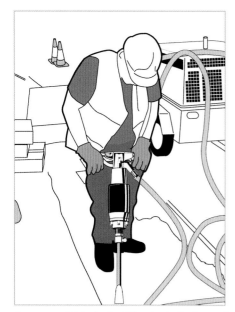

Compressed air tools should be checked before and after use

Non-powered hand tools

These may seem low risk, but are responsible for many injuries; they need to be well maintained and regularly inspected. Well used chisels and bolsters can form mushroom heads. When they are struck, fragments can fly into the air and into the eye. Loose handles, blunt blades and worn parts all pose a risk.

Lasers

If used correctly lasers should not pose a health hazard.

A rotating laser means it is difficult to look directly at the beam for more than an instant. Static lasers (such as pipe lasers), pose more of a risk.

Exclusion zones and warning signs must be in place if high-powered lasers are being used.

 For further information refer to GE 700 *Construction site safety* **Chapter C05 Plant and work equipment.**

18
Site transport safety

What your employer should do for you	246
What you should do as a supervisor	247
Introduction	248
Planning for safety	248
Pedestrian and vehicle segregation	248
Deliveries, loading and storage areas, and distribution	250
Reversing vehicles	251
Perimeters, interface with highways and public rights of way	252

Site transport safety

	What your employer should do for you
1.	Plan to minimise the risk of the people/vehicle interface.
2.	Organise for separate pedestrian routes.
3.	Ensure that statutory records are kept up to date and maintain registers.
4.	Implement procedures for the safe delivery, off-loading and movement of plant and materials onto and around the site.
5.	Ensure that drivers are instructed in permitted situations for reversing vehicles.
6.	Maintain up-to-date lists of authorised drivers and operators.
7.	Plan vehicle movements to minimise the risks at access and egress locations.
8.	Monitor that movement of plant is in line with the traffic management plan.
9.	Ensure that traffic routes are observed by site vehicles as applicable.
10.	Monitor that operators of mobile or self-propelled plant have adequate visibility in all directions and all visibility aids are in good order.

C18

Site transport safety

What you should do as a supervisor

Checklist		Yes	No	N/A
1.	Supervise the site to minimise the risk of the people/vehicle interface.			
2.	Monitor and maintain separate pedestrian routes.			
3.	Maintain statutory records and keep registers up to date.			
4.	Liaise with managers for the safe delivery, off-loading and movement of plant and materials onto and around the site.			
5.	Check that drivers are instructed in permitted situations for reversing vehicles and provide a signaller.			
6.	Ensure that only authorised drivers and operators move and operate plant.			
7.	Organise vehicle movements to minimise the risks at access and egress gates.			
8.	Follow the traffic management plan for vehicle movements, loading, off-loading and storage.			
9.	Manage traffic routes to ensure they are observed and followed by site vehicles.			
10.	Check that operators of mobile or self-propelled plant have adequate visibility in all directions and all visibility aids are in good order.			

C18

Site transport safety

Introduction

Each year within the construction industry there are hundreds of preventable accidents and injuries. This not only affects the lives of workers but can also result in material damage.

Accidents occur throughout the construction process, from groundworks to finishing trades. Managers, supervisors, workers, visitors to sites and members of the public can all be at risk if construction vehicle activities are not properly managed and controlled.

The majority of construction transport accidents result from the inadequate segregation of pedestrians and vehicles. This can usually be avoided by careful planning, particularly at the design stage, and by controlling vehicle operations during construction work. Inadequate planning and control is the root cause of many construction vehicle accidents.

Planning for safety

Principal and main contractors should ensure that pedestrians and vehicles are adequately separated by establishing a site traffic management plan typically including:

- ✓ pedestrian-only areas from which vehicles are completely excluded
- ✓ safe designated pedestrian routes to work locations
- ✓ vehicle-only areas, especially where space is limited or traffic is heavy
- ✓ safe vehicle routes around the site
- ✓ traffic routes/haul roads that must be kept clear of obstructions.

Pedestrian and vehicle segregation

Where work space permits, establish pedestrian routes on site that provide safe pedestrian access to work areas. Pedestrian routes should be either located a safe distance away from areas of vehicle activity or provided with appropriate physical protection (such as substantial fencing and/or kerbs), to prevent pedestrians being struck by vehicles or their loads. Pedestrian routes should:

- ✓ be clearly separated from vehicle routes by fencing and/or a kerb, or other suitable means
- ✓ be wide enough to safely accommodate the number of people likely to use them at peak times
- ✓ allow easy access to work areas
- ✓ be kept free from obstructions and tripping hazards and be clearly signed
- ✓ ensure pedestrian safety where they cross main vehicle routes
- ✓ provide pedestrians with a clear view of traffic movements at crossings and where gates used by pedestrians lead onto traffic routes
- ✓ have clearly marked, separate access for pedestrian use at loading bays and site gates used regularly by construction vehicles
- ✓ be adequately lit where light is not provided by natural means.

Site transport safety

The safe separation of construction plant and pedestrians avoids accidents

C18

Site transport safety

Deliveries, loading and storage areas, and distribution

In some situations it may not be possible to phase certain jobs to keep plant and pedestrians apart so there is a greater chance of an accident occurring. Where it is inevitable that plant and people will be operating in close proximity (for example, a telehandler delivering a pack of bricks for the bricklaying gang), it is essential that:

- [✓] everyone involved is informed of and fully appreciates the potential for danger and keeps out of the danger area
- [✓] the plant operator is fully competent
- [✓] the activity is under the control of one person only who instructs the others what to do
- [✓] the period of risk is kept to a minimum.

Work on site should be planned to minimise vehicle movements, and to avoid unnecessary deliveries and the double handling of materials on site. The location of loading and storage areas needs to be carefully considered. Where there is little on-site storage space, off-site storage areas may be required for the temporary storage of materials. Loading and storage areas should:

- [✓] be located away from pedestrian-only areas and main pedestrian routes
- [✓] exclude pedestrians so far as reasonably practicable
- [✓] have one-way systems and safe exit points
- [✓] have sufficient room for vehicle movements
- [✓] have adequate fixed lighting, signs and appropriate visibility aids for drivers (for example, convex mirrors positioned on corners).

Load and unload vehicles at level ground and avoid climbing onto the vehicle unless fall prevention/arrest measures are provided, in areas away from passing traffic, pedestrians and overhead hazards (for example, bridges, pipelines or electrical cables). Loads need to be:

- [✓] of suitable height and width for the vehicle and road conditions on site
- [✓] secured to prevent movement
- [✓] evenly loaded and distributed to keep the centre of gravity as low as possible and to prevent stresses on vehicle structures
- [✓] positioned on vehicles and transported so that they do not adversely affect vehicle stability
- [✓] checked to ensure they will not fall uncontrollably when restraints are removed during unloading.

No vehicle should be loaded beyond its safe working capacity. Loads that project out from the body of the vehicle should be indicated by a warning beacon/flag or sign.

Segregated parking, storage and access areas on a large site

Site transport safety

The precautions necessary for managing the risks presented by hazardous loads when transported by road need to be followed on site and supplemented as necessary in relation to site risks (for example, lorries carrying LPG cylinders should not be parked near scaffolds where there is a risk of falling objects striking them).

Site rules should require visiting drivers to inform site management of any hazardous loads on their vehicles. Appropriate fire precautions need to be instituted for loads that contain substances with specific fire hazards (such as fuels and solvents). Information about the hazards of dangerous loads and necessary precautions in the event of an accident should be issued to all site drivers.

Where vehicles are transported on site on low-loaders, they should be:

- ☑ dismantled so far as possible to keep them within the dimensions of the carrying vehicle
- ☑ emptied of fuel, as much as possible
- ☑ relieved of hydraulic pressure by moving all control levers through all positions, twice, before transportation
- ☑ secured and restrained to prevent movement with their parking brake applied and wheels and rollers chocked.

Loading and off-loading areas should be of sufficient size to allow vehicles to move, without striking obstructions or causing hazards to others. Access ramps used for getting vehicles on and off low-loaders should be of adequate strength and size.

Reversing vehicles

Vehicle reversing operations cause a third of all fatal transport accidents in the construction industry, producing an average of five deaths and 20 major injuries per year. The most effective way of managing the risks from reversing is to avoid the need for reversing manoeuvres by providing one-way systems, turning areas and drive-through loading and unloading areas.

When planning and controlling site vehicle operations, the hierarchy of control measures for reversing operations is shown below.

1.	Eliminate the need to reverse.
2.	Reduce reversing operations.
3.	Segregate vehicles and pedestrians.
4.	Ensure safe systems of work are followed.
5.	Provide warnings when vehicles are reversing.

Vehicles required to reverse on site should provide adequate visibility around the vehicle for the driver to ensure safety. Safe systems of work need to be devised and followed for all reversing operations, particularly when signallers are used to control third-party risks or assist in the accurate positioning of the vehicle.

Warning systems offer the lowest level of protection in the hierarchy and, if they are the only precaution used, are only appropriate for low-risk situations.

Site transport safety

Perimeters, interface with highways and public rights of way

For most sites the perimeter is a geographical area within which construction work will be carried out.

Identifying and establishing this perimeter is an important aspect of managing public risk.

Three issues should be considered when deciding on the type of perimeter guarding:

- ☑ **planning** what form the perimeter will take (solid hoarding, fence panels and so on)
- ☑ **providing** the perimeter
- ☑ **maintaining** what has been provided.

Sometimes construction work can create risks to the public outside the site perimeter (for example, unloading materials from a delivery lorry outside the perimeter).

These risks might include materials falling from access platforms, materials stored temporarily off site, the operation of cranes and other lifting equipment either on or off site, vehicles blocking the footway, obstruction of the roadway, obscuring lines of sight and causing congestion.

 The situation can change as construction work progresses and must be monitored and controlled continuously.

The site entrance is an interface with the site and the highway, the general public and their rights of way, and with road traffic, and has to be planned and managed as a matter of priority to remove the risk of accident and incident as far as possible. Site entrances must be clearly marked by signs and there should be separate access for vehicles and pedestrians. Vehicle access and egress must be closely controlled.

There will be a need to liaise with the local highways authority, and you may need authorisation from them if your work involves the closure or obstruction of public footpaths or roads. You will probably need a licence from them before you can begin. The licence may set standards that describe aspects of how, for example, a scaffold should be constructed, how it should be marked (paint, tape and so on), and when it needs to be lit. But regardless of this, there are always certain precautions that you can take:

- ☑ exclude the public from the work area whenever possible
- ☑ fence off the area and provide alternative routes which are clearly signposted
- ☑ erect, modify and dismantle equipment when there will be fewer members of the public in the area and always use warning notices
- ☑ fans, tunnels and sheeting are a useful means of protection (make sure the scaffold is designed to take the extra loading and wind resistance)
- ☑ ask for protective measures to be put in place at an early stage during erection and have them removed as late as possible during dismantling
- ☑ lighting may be necessary in tunnels
- ☑ use brick guards, netting or other suitable protection to prevent materials falling
- ☑ do not drop or throw components during erection or dismantling

Site transport safety

- [✓] make sure the working platform is constructed to prevent materials falling through it; double board scaffold platforms and insert a layer of strong polythene between the two sets of boards (a few small punctures will allow rainwater to drain away)

- [✓] make sure scaffold components do not project where there is a risk to people or vehicles

- [✓] bolts on couplings should face away from the public or be wrapped

- [✓] consider enclosing the base of the scaffolding to prevent climbing, especially on or near occupied residential premises and schools

- [✓] out of hours, remove ladders from the scaffold; secure them in a compound or in storage containers

- [✓] make sure that doors to buildings or those allowing access to the roof, lifts, motor rooms and so on are locked at all times when work is not in progress

- [✓] consider using alternatives to scaffolding (such as mobile elevating work platforms, cradles and mast climbers).

For further information refer to the HSE's *The safe use of vehicles on construction sites* (HSG144) and *Protecting the public – your next move* (HSG151).

Site transport safety

C 18

19
Lifting operations

What your employer should do for you	256
What you should do as a supervisor	257
Legislation	258
Definitions	258
Competence in lifting operations	258
Safe system of work	258
Cranes	260
Overhead power cables	261
Excavators used for lifting	262
Telescopic materials handlers	263
Gin wheels	264
Maintenance and inspection	265
Tower cranes	266
Information and instructions	267

Lifting operations

What your employer should do for you

1. Put procedures in place to ensure that personnel are trained, competent and appointed.
2. Ensure that equipment in use is suitably inspected, tested and thoroughly examined.
3. Provide lifting plans and ensure that the work team are briefed.
4. Ensure that lifting equipment and accessories are marked and are within their examination date.
5. Survey the site and implement checks for underground services, voids and overhead lines.
6. Ensure that there are no restrictions on airspace or oversailing rights.
7. Put procedures in place to prevent the fall of anyone involved in the off-loading of vehicles.
8. Ensure that if excavators, loaders or combined excavator loaders are used, they have the required accessories and documentation.

Lifting operations

What you should do as a supervisor

Checklist	Yes	No	N/A
1. Follow procedures put in place, to ensure that personnel are trained, competent and appointed for the work in hand.			
2. Check that equipment in use is suitably inspected, tested and thoroughly examined.			
3. Follow lifting plan requirements and ensure that the work team are briefed and follow the plan.			
4. Check that lifting equipment and accessories are marked and are within their examination date and reject non-complying equipment.			
5. Ensure that the survey of the site is accurate and implement checks for underground services, voids and overhead lines.			
6. Follow employer guidance on restrictions on airspace or oversailing rights and make the work team aware of any restrictions.			
7. Plan work and put measures in place to prevent the fall of anyone involved in the off-loading of vehicles.			
8. Check that if excavators, loaders or combined excavator loaders are used, they have the required accessories and documentation.			

C19

Legislation

The Lifting Operations and Lifting Equipment Regulations cover all aspects of mechanical lifting operations. They are accompanied by an Approved Code of Practice and guidance notes. Any employer who carries out any mechanical lifting operations using lifting equipment must comply with the requirements of the regulations.

Because lifting equipment is also work equipment under the Provision and Use of Work Equipment Regulations, these regulations are also relevant to lifting operations. These regulations consider lifting equipment with regard to such factors as suitability, maintenance, keeping in good order and the use of trained and competent operators.

For further information on work equipment refer to Chapter C17 Work equipment and hand-held tools.

Definitions

The regulations cover the safety aspects of using lifting equipment and lifting accessories, including planning.

Lifting equipment is any work equipment (such as cranes (mobile and static), hoists, telehandlers and excavators) that is used for mechanically lifting or lowering any load, including people.

The definition includes:

- ☑ all attachments used for anchoring, fixing or supporting the lifting equipment
- ☑ hand-operated lifting equipment (such as gin wheels and hand-operated hoists).

Lifting accessories are items of equipment used for attaching the load to the lifting equipment (such as chains, ropes, slings, hooks, shackles, spreader-beams and eye bolts).

The regulations do not apply to shovels, crowbars or wheelbarrows or anything else that would be regarded as equipment for assisting the manual handling of loads.

Competence in lifting operations

Where it is necessary to hire in lifting equipment, the employer must decide whether there is adequate competency available within the company to plan, organise and perform the lift(s). If this is not the case it will be necessary for the hirer to arrange for the crane hire company to appoint competent people and take all legal responsibility for the lift(s). This option is known as contract lifting.

Safe system of work

Principles

Inadequately planned lifting operations are often responsible for accidents that result in fatalities and injuries to those involved and/or the structural failure or overturning of the lifting equipment.

Employers have a duty to ensure that each lifting operation is:

- ☑ properly planned by a competent person (referred to as the appointed person who's duty is to develop/approve the lifting plan)
- ☑ adequately supervised
- ☑ carried out in a safe manner.

Lifting operations

Each lifting operation should be the subject of a risk assessment, although where a lifting operation is repeated in the same circumstances with the same hazards and level of risk (such as stacking materials using a forklift truck), an initial, generic risk assessment should satisfy the legal requirements for a suitable and sufficient risk assessment.

The findings of the risk assessment should be used as a basis for developing the safe system of work. Where a lifting operation is not straightforward, it might be necessary to write a method statement (lifting plan) for the job, which in effect will become the safe system of work, formally written down.

The safe system of work should be effectively communicated to all those involved in the lifting operation. It must include:

- ☑ thorough planning of the operation, along with the selection, provision and use of suitable lifting equipment and accessories
- ☑ a requirement that all equipment must be maintained, tested and examined as necessary
- ☑ the provision of all test certificates and other documentation relevant to the equipment being used
- ☑ a requirement that all equipment is operated by trained and competent people
- ☑ supervision by trained and competent staff, with authority to stop a job if necessary
- ☑ the prevention of unauthorised use or movement of equipment by suitable security measures
- ☑ the safety of all persons, both those involved in the lift as well as those not involved but who may be affected by the lifting operation.

For further information refer to Chapter A04 Risk assessments and method statements.

Lifting operations, no matter how large or small must be properly planned

Failure

Accidents with cranes and other lifting equipment are often caused by:

- ☑ using lifting equipment of the wrong type or lifting capacity
- ☑ using lifting equipment in an incorrect manner (unsafe technique)
- ☑ incorrect slinging of the load
- ☑ lifting a load of an unknown and underestimated weight
- ☑ poor maintenance of the equipment
- ☑ undue haste carrying out the task

Lifting operations

- ☑ failure of the ground supporting the lifting equipment
- ☑ lack of training of the crane operators and signallers
- ☑ lack of co-ordination of crane movements when more than one crane is operating and their arcs of movement overlap.

 The collapse, overturning or failure of many items of lifting equipment is a notifiable dangerous occurrence, which must be reported to the HSE, even if no person is injured.

Cranes

Siting

The siting of cranes will usually be carried out by the site management in conjunction with the crane operator, who must ensure that:

- ☑ the position corresponds to those approved in any lifting plan
- ☑ the stability of the crane will not be adversely affected by unstable ground conditions, or the presence of excavations, manholes, cellars, or other underground voids
- ☑ safe access is available to refuelling or other service vehicles
- ☑ the crane does not encroach into the safety distance from overhead cables (see next page)
- ☑ oversailing rights must be agreed (tower cranes mainly)
- ☑ airspace restrictions must be checked (tower cranes mainly)
- ☑ suitable clearance (minimum 600 mm) is maintained between any mobile or slewing crane and any fixture, in other words, guard-rail, adjacent building and so on (there is no crush zone).

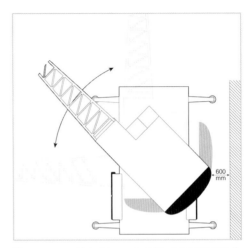

Clearance must take into account the reduced space if the crane tips

Assistance

Lifting operations involving the use of a crane will usually require the assistance of a signaller/slinger to safely secure the load to the crane and give direction to the operator. An effective system of communication between them is essential – usually hand signals, which may be supplemented by radio content.

Signals

The giving of signals by untrained persons is extremely dangerous and must be prohibited. Where hand signals are to be used, they should be those that are detailed in BS 7121-1, or the Health and Safety (Safety Signs and Signals) Regulations.

Lifting operations

The crane operator and signaller must ensure that they know which set of signals is being used, as there are differences between the two sets. Care must be exercised by those giving and receiving signals and, if there is any doubt, movement of the load must be stopped and the signal repeated if necessary.

Drivers

Drivers of cranes and other lifting appliances, and any other people involved in lifting operations, including signallers, must be trained, experienced and (it is recommended) aged 18 years or over unless under the direct supervision of a competent person for the purposes of training.

Overhead power cables

The use of cranes and other lifting appliances near overhead power cables calls for extreme care. Particular care must be taken with the positioning of cranes, piling rigs, hiabs (lorry loader cranes), tipper trucks and so on. The higher the voltage in the cables, the greater the safety distance that is required.

> The power supply company must always be consulted before siting any plant adjacent to overhead power cables. They will advise on the voltage of the supply and the minimum safe distance.

> If in any doubt always ask.

As a guide, the minimum safe working distance between the lifting equipment (usually the tip of the jib) and the power cable is:

- ☑ 9 m if the cables are suspended on wooden poles
- ☑ 15 m if the cables are suspended on metal pylons.

(Both measurements should be taken from a point on the ground directly below the outermost cable to the tip of the jib when it is nearest to the cable.)

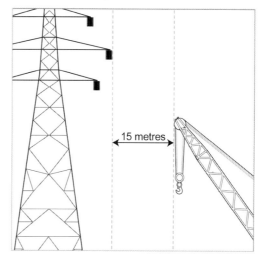

Clearance of 15 m between the tip of the jib and the outermost cable

In circumstances where it is not possible to maintain the minimum safety distance, the electricity supply company must be consulted to establish a safe system of working. Ideally, the supply will be isolated by the supply company for the duration of the work.

Lifting operations

Danger areas should be staked out and flags put up.

Cranes and machines should only pass under live cables where goal posts have been set up, as specified in HSE Guidance Note GS6 or by the local electricity supply company.

Other obstructions

Some modern machines are equipped with sensors that limit the height to which the jib will reach and the limit to which a machine can slew. Warning lights and an alarm indicate when the machine is about to exceed the limit (alarm plus amber light) and when it has been exceeded (alarm plus red light).

Loads

The safe working load (SWL) of the appliance should be indicated on all cranes and lifting appliances and must never be exceeded. Rated capacity indicators (previously known as automatic safe working load indicators) are provided and should never be interfered with.

The SWL includes the weights of the hook block and all attached lifting gear. These weights must be deducted from the listed SWL when calculating the net load.

When lifting near to the maximum SWL, the load should be raised a short distance and then stopped to check stability and safety.

Excavators used for lifting

Excavators are mobile work equipment and are therefore subject to the requirements and provisions of the Provision and Use of Work Equipment Regulations. They are particularly subject to the requirements contained in Part II, Regulations 4 to 24 and Part III, in particular Regulations 25 and 26.

Excavators, loaders and combined excavator loaders may be used as cranes, in connection with work directly associated with an excavation, and any other application where this type of equipment has been designed for the purpose and can be used safely.

All work is subject to a suitable and sufficient risk assessment, subsequent control measures and the capabilities of the work equipment.

The risk assessment should take into account that when a machine is in the object handling mode (being used as a crane), it will be necessary for the slinger to approach the machine to hook the load on and off. This person will be in what is regarded as a hazardous area and much nearer to the machine than anyone would be in normal circumstances. The slinger is at risk of being struck by the load, bucket or excavator arm if the excavator moves or slews rapidly. Excavator operators and slingers must be made aware of these dangers; effective communication and constant vigilance are essential.

The risk assessment must also establish whether the machine is suitable for the proposed task. The weight of the bucket (if still fitted), plus the quick hitch must be added to the weight of the load to establish if the machine will be working within its safe working load. Ideally, unless there are good reasons for not doing so, the bucket should be removed to improve the machine operator's visibility.

The risk assessment must also address the:

- ☑ need for the lifting operation to be ideally segregated from other work activities taking place in the vicinity, particularly where it is necessary for the machine to travel with a raised load

Lifting operations

- ☑ ground conditions, particularly where a tracked excavator will carry out the lifting operation. Such machines have no means of levelling themselves and are therefore dependent upon the ground being sufficiently level to track across it and carry out the lifting operation safely.

The safe working load must be clearly marked on the machine and any lifting accessories (such as a quick hitch).

 A rated object handling capacity table must be available in the cab.

If the rated lifting capacity for an excavator (or the backhoe of a backhoe-loader) is greater than 1 tonne (or the overturning moment is greater than 40,000 Nm), the machine must be fitted with:

- ☑ a boom lowering control device on the raising boom cylinder(s) (a safety check valve), which meets the requirements of BS 6912-1
- ☑ an acoustic or visual warning device, which indicates to the operator when the object handling capacity or corresponding load moment is reached.

Chains or slings for lifting must not be placed around or on the teeth of the bucket. Accessories for lifting may only be attached to a purpose-made point on the machine.

Rated capacity indicators (RCI) are particularly useful where an excavator, being used as a crane, must slew as part of the operation. The RCI will continually calculate whether the machine is within its safe lifting capacity for varying configurations of radius and jib angle.

A proprietary manhole lifting attachment

Telescopic materials handlers

Known as telehandlers, these are commonplace on construction sites. The range and type varies considerably and include rough terrain, two or four wheel drive and crab or articulated steering. The ability of telehandlers to raise loads to greater heights increases the hazard of overturning. To counteract this, some types are fitted with stabilisers, outriggers or chassis-levelling devices.

Lifting operations

Many machines are used with a variety of attachments that may affect their stability. The manufacturer's or authorised supplier's recommendations for fitting and using attachments should be followed.

Operators and supervisors should be aware that attachments will alter the rated lift capacity and centre of gravity of the machine.

Machines should normally be fitted with rated capacity indicators (RCIs). These give warning of approaching overload and should always be switched on during load handling operations. RCIs must be maintained and tested according to the plant's instructions as part of the inspection regime required under PUWER.

Telehandlers should have, and follow, a lifting plan

A telehandler should have a safe system of work (lifting plan), which is specific to the site conditions, loads and attachments.

Attachments (such as tipping slips) should be regularly inspected and checks made to ensure restraining chains are secure and being used.

 For more information refer to the HSE's ACoP and guidance, operator training and safe use of rider-operated lift trucks.

Gin wheels

Often, provision has to be made to raise tools and light materials manually up to a working platform. The following provisions and requirements apply to gin wheels supported by a scaffold.

- ☑ Poles and hooks should be strong enough to take the load that is to be lifted and be properly secured to prevent movement.

- ☑ All ropes should comply with the relevant BS EN standards and fit the wheel correctly. They should be marked with a tag confirming their safe working load.

- ☑ Proper hooks for suspending the gin wheel from the scaffold should be used, preferably of the ring type with a swivel eye to fit over the tube. They should be secured by fittings to prevent any lateral movement.

- ☑ Any joints in standards should be made with sleeve couplers.

- ☑ Gin wheels should be suspended not more than 750 mm from the outer support.

- ☑ Hooks used for raising the load should be safety hooks and spliced into the rope.

- ☑ The maximum loading should be no more than 50 kg at 750 mm from the outer support.

- ☑ Gin wheels must be visually checked on a regular basis, thoroughly examined before use and every 12 months and a certificate issued.

Lifting operations

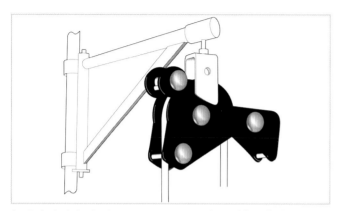

Inertia braked gin wheels are now common practice and far safer than traditional gin wheels

Maintenance and inspection

The Lifting Operations and Lifting Equipment Regulations require all lifting equipment to be thoroughly examined by a competent person.

Definition of competent person

A competent person is one who has the appropriate practical and theoretical knowledge and experience of lifting equipment that will enable them to carry out a thorough examination or inspection, to detect defects or weaknesses and assess the importance of such defects or weaknesses in relation to the safety and continued use of the lifting equipment.

If an owner does not have the necessary competence in house, these examinations are often carried out by a representative of the insurance company.

Thorough examinations

Lifting equipment must be thoroughly examined:

- ☑ when it is first used (unless bought brand new)
- ☑ if it is installed, after installation but before use
- ☑ if it is assembled, after assembly but before use
- ☑ at intervals not exceeding 12 months.*

* Lifting equipment used for lifting persons, and all lifting accessories must be thoroughly examined every six months.

All thorough examinations should be conducted by a competent person in accordance with an examination schedule drawn up by a competent person. Reports of thorough examinations must be prepared and supplied to the employer.

The competent person who carries out thorough examinations is unlikely to be the same competent person who plans the lift.

Inspections

Where a risk assessment under the Management Regulations has identified risks to operators or other persons that would be addressed by regularly inspecting the lifting equipment, the employer must ensure that such inspections are carried out by a competent person.

A competent person must inspect all lifting appliances at suitable intervals (usually before first use and then weekly). The frequency and extent of the inspections will depend on the potential for failure of the lifting equipment. The inspection should include visual checks and functional tests.

Lifting operations

Report of thorough examinations

For every thorough examination, the competent person must record the following information:

- ☑ the name and address of the employer
- ☑ the address of the premises
- ☑ the identification of the equipment
- ☑ the date of last thorough examination
- ☑ the SWL of the equipment
- ☑ whether it is a first examination, or six-monthly, or 12-monthly, or an examination under an examination scheme
- ☑ any defects found
- ☑ particulars of any repairs or alterations needed to remedy the defect, and the date by which the next thorough examination must be carried out
- ☑ any testing necessary at the next thorough examination
- ☑ the name, address and qualifications of the competent person carrying out the thorough examination
- ☑ the signature of the person and date of the thorough examination

Records can be kept in writing, or electronically on a computer.

Tower cranes

A conventional tower crane is defined in the regulations as a slewing jib type crane with its jib located at the top of a vertical tower and which is assembled on a construction site from components. This includes, but is not limited to, cranes with horizontal or luffing jibs and slewing rings at the base or top of the tower. These tower cranes are usually installed (and dismantled) with the assistance of another crane and, as a result, are sometimes referred to as assisted erected cranes.

There are four key aspects to the safe use of cranes. These are planning lifting operations, safe systems of work, supervision of lifting and thorough examination.

The duty to notify the HSE of the erection of a conventional tower crane in accordance with the Notification of Conventional Tower Crane Regulations has now been revoked, and the associated register ceased from 6 April 2013.

The Construction Plant-hire Association (CPA) have produced a series of best practice guides that give detailed guidance for supervisors and managers, on subjects such as risk assessment, thorough examination of lifting equipment and safe use and maintenance of plant.

Topics covered include the competence of those erecting and dismantling tower cranes; thorough examination, inspection and maintenance of tower cranes; and the management of the installation and dismantling process. Publications are available as free downloads from the CPA's website at www.cpa.uk.net

Lifting operations

Where several tower cranes are close together, their operation must be co-ordinated to avoid collisions

Information and instructions

There is a general requirement in health and safety law for employers to ensure that employees, including supervisors and managers where appropriate, are provided with adequate information, instruction, training and supervision. This principle must be applied to the use of lifting equipment, although the level of information, instruction and so on, provided should be proportional to the complexity of the lifting equipment being used. The manufacturer's instructions will be a source of useful information.

The information should include:

- ☑ how and when the equipment may be used
- ☑ foreseeable abnormal conditions and the action to be taken should such conditions arise
- ☑ conclusions drawn from experience of using the equipment.

The information and instructions must be easily understood.

The Provision and Use of Work Equipment Regulations also require that:

> Every employer shall ensure that, where any machinery has a maintenance log, the log is kept up to date.

For further information refer to GE 700 *Construction site safety,* Chapter C07 Lifting operations and Chapter C08 Lifting equipment.

Lifting operations

C19

20

Working at height

What your employer should do for you	270
What you should do as a supervisor	271
Introduction	272
Legislation	272
Hierarchy for working at height	273
Preventing falls	274
Scaffolding	275
Mobile tower scaffolds and access platforms	279
Podium steps	280
Ladders and stepladders	281
Trestles and lightweight staging	283
Mobile elevating work platforms	283
Roof work	287
Fragile roofs	290
Fall-arrest systems	292
Safety harnesses and lanyards	293
Working over or near to water	294
Working above other people	295

Working at height

	What your employer should do for you
1.	Assess any elements of the task that minimise work at height.
2.	In all cases work at height will be carried out using the most appropriate means of access.
3.	Avoid workers going on fragile roofs by replacing profiled roof sheets or roof lights from underneath. If this is not possible, a safe system of work must be planned and followed.
4.	Make arrangements for scaffolds to be erected, altered and dismantled by a competent person.
5.	Arrange for the statutory inspection of scaffolds and plant to be carried out by a competent person.
6.	Ensure that all work at height is planned, supervised and carried out by a competent workforce.
7.	Provide training to ensure that competent operators will be available to operate mobile elevating work platforms (MEWPs).
8.	Check that ground conditions are suitable for the use of MEWPs and the erection of scaffolds, including investigating the existence of cellars, drains and other underground voids.
9.	Survey the surrounding ground for the possible existence of overhead power lines and other high-level hazards.
10.	Restrict the use of ladders to tasks where it is not reasonably practicable to use an alternative, safer means of access.
11.	Ensure that adequate fall prevention systems are in place for all work at height activities.

D20

Working at height

What you should do as a supervisor

Checklist	Yes	No	N/A
1. Plan and supervise any elements of the task that minimise work at height.			
2. Ensure that in all cases work at height will be carried out using the most appropriate means of access.			
3. Ensure that working on fragile roofs is avoided and adequate arrangements are put in place to provide a safe place of work.			
4. Check the arrangements for scaffolds to be erected, altered and dismantled by a competent person.			
5. Ensure that the statutory inspection of scaffolds and plant is carried out by a competent person.			
6. Supervise all work at height and that it is carried out by a competent workforce.			
7. Check to ensure that operators are competent to operate mobile elevating work platforms (MEWPs).			
8. Monitor ground conditions to ensure that they are suitable for the use of MEWPs and the erection of scaffolds and carry out investigations for the existence of cellars, drains and other underground voids.			
9. Establish safe systems of work to avoid overhead power lines and other high-level hazards.			
10. Monitor the use of ladders to ensure that the safest means of access is provided.			
11. Monitor all working at height activities.			

D20

Working at height

Introduction

Falls from height continue to be the main cause of fatalities and specified injuries within the construction industry. Falls that cause injuries do not always occur from a great height. Death and major injuries have resulted on many occasions from falls from less than 2 m above the ground.

 Working at height should be regarded as working in any place from where a fall could cause personal injury.

Falls resulting from the use of ladders, working on or near to roof edges or fragile materials continue to figure prominently amongst the accident statistics. A risk assessment must be carried out before performing any work at height.

Legislation

All work at height must be carried out in compliance with the **Work at Height Regulations**. Employers, managers and supervisors should note that the key provisions of the regulations are:

- ☑ where it is reasonably practicable, avoid the need to carry out work at height
- ☑ where work at height cannot be avoided, select the most appropriate work equipment for the work
- ☑ ensure that the way that work is carried out is based upon the findings of a risk assessment
- ☑ as far as reasonably practicable, organise work at height so as to prevent falls and falling objects (such as tools and materials)
- ☑ reduce the distance of and potential consequences of any fall that does occur
- ☑ give priority to the use of passive fall-protection measures (such as safety nets), over those (such as safety harnesses), which the wearer has to clip on
- ☑ ensure that those who have to work at height are competent and fit to do so
- ☑ the height to which a scaffold can be built, before it is necessary to install guard-rails and toe-boards, must be decided by the findings of a risk assessment, which must consider the minimum height from which a fall would be likely to result in an injury
- ☑ the minimum height of the top guard-rail is 950 mm and the maximum gap between the intermediate rail and both the top guard-rail and the top of the toe-board is 470 mm
- ☑ ladders and stepladders must only be used where it can be shown, through a risk assessment, that it is not reasonably practicable to use safer types of access equipment and the remaining risks are low
- ☑ each working platform must be wide enough for its intended use; BS EN 12811 recommends a minimum width of 600 mm
- ☑ there is no minimum height for a toe-board since it is usual to use scaffold boards on edge, although the above reference recommends a minimum height of 150 mm.

Working at height

Hierarchy for working at height

> **Step 1 – Avoid working at height**
> e.g. assemble on the ground and crane up, use extendable handles on equipment
> Only if this can't be avoided then:

> **Step 2 – Use an existing safe place of work**
> e.g. parapet walls, defined access points, staircases
> Only if this can't be done then:

> **Step 3 – Provide work equipment to prevent falls**
> e.g. scaffolding, edge protection, handrails, podiums, mobile towers, MEWPs
> Only if this can't be provided then:

> **Step 4 – Mitigate distance and consequence of a fall**
> e.g. safety netting, airbags, crash decks
> (harnesses must be used as a last resort)

> **Step 5 – Provide instruction, training and/or other means**
> e.g. PASMA training for mobile towers, IPAF training for MEWPs
> using ladders (the last resort)

Working at height

Preventing falls

Ideally, falls should be prevented by physical barriers and equipment. Methods include scaffolding, mobile towers, MEWPs, suitably selected podiums, edge protection systems, and netting, brick guards or solid boards should be used to prevent falls of materials. Plastic barriers, netting or rope and pins are not suitable as edge protection to prevent people from falling.

Proprietary edge protection fixed to steel at ground level prior to lifting into position

Proprietary edge protection and triple guard-rails in place

Holes in floors, slabs, roofs and open shafts

Secure and fix load-bearing covers with warning signage, or erect securely anchored double guard-rails complete with toe-boards, where holes have been cast or cut in floor slabs, roofs and so on, and around open stairwells, service voids and lift shafts.

Working at height

Proprietary edge protection

Scaffolding

Competence
The design, erection, alteration and dismantling of scaffold should only be done by competent workers, or be carried out under the supervision and direction of competent persons. It is an offence for anyone else to erect, modify or dismantle a scaffold. All but standard scaffolds must be designed by a competent scaffold designer.

 For further information refer to the HSE's website at www.hse.gov.uk/construction/safetytopics/workingatheight.htm and to the National Access and Scaffolding Confederation's website at www.nasc.org.uk

Inspections and reports
Scaffolds must be periodically inspected by a competent person. Scaffolds must not be used unless they have been inspected:

- ☑ before being used for the first time
- ☑ at intervals not exceeding seven days
- ☑ after adverse weather conditions or any other event likely to have affected the scaffold
- ☑ after any modification or alteration.

Inspection tag on scaffold

Working at height

 All places of work at height must be checked to ensure they are safe to work from, prior to being used on each occasion.

Reports of inspections and checks of places of work at height are required under Regulations 12 and 13 of the Work at Height Regulations, respectively.

 Where a scaffold will be erected for you by a scaffolding contractor, you should check that they will comply with NASC Guidance Notes TG20 and SG4: and that you will receive a copy of their handover certificate that certifies that the structure is complete and it has been correctly erected for the purpose (loading) for which it has been designed.

Safety requirements for scaffolding, in general, are shown below.

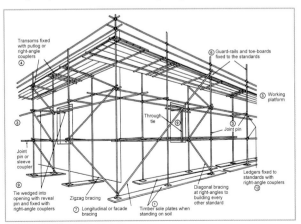

Foundations should be level and firm, using base plates and sole boards where necessary, with the following dimensions:

- ☑ hard ground: 450 x 225 x 35 mm
- ☑ soft ground: 760 x 225 x 35 mm.

Standards are vertical scaffold poles and should be spaced closely enough to provide an adequate support for the proposed load.

Ledgers are horizontal scaffold poles, linking the standards along the scaffold.

Transoms are horizontal scaffold poles running from front to back, supporting the scaffold boards that form the working platform.

Ties: apart from special cases (such as bird-cage scaffolds), other scaffolds must be securely tied to the structure against which they are erected. The number and positioning of the ties will be determined by the scaffold designer.

Bracing is the method by which scaffold structures are made rigid.

Facade bracing is achieved by securing scaffolding tubes across the face of the scaffold parallel to the face of the structure. The bracing tubes may be a continuous diagonal from the top to the bottom of the scaffolding, or may be installed in a zigzag configuration.

Ledger bracing and plan bracing are other means by which a scaffold structure may be stiffened.

Platforms must be wide enough to provide a safe place of work and enable other people and equipment to pass. British Standard BS EN 12811-1 recommends a minimum width of 600 mm.

Stability of boards: there should be at least three supports for each scaffold board, normally not more than 1.5 m apart. Boards should overhang each end support by at least 50 mm but by not more than four times the thickness of the board.

Guard-rails should be placed along the outside edges and at the ends of each working platform from where a person could fall and suffer personal injury, as shown in a risk assessment. Occasionally, guard-rails are necessary on the inner edge of working platforms.

In most circumstances a minimum of two guard-rails should be installed. The top guard-rail should be at least 950 mm above the working platform and the other mid-way between the top guard-rail and the top of the toe-board. There should be no unprotected gap of more than 470 mm. A substantial barrier that cannot be displaced may be fitted as an alternative to the mid guard-rail, but it must have fall-protection properties equal to or better than a mid guard-rail.

 If a guard-rail has to be moved, it must only be done with permission and only by a trained and competent person. The guard-rail must be replaced as soon as possible.

Access: ideally access to a scaffold will be via a purpose-built stair tower. However, where the decision is taken to use a ladder, the scaffold should be designed to incorporate internal ladders. If an external ladder is used it should be positioned along the face of the scaffold rather than at right-angles to it. Ladder gates should be installed at access points.

A purpose-built scaffold stair tower

Working at height

A ladder safety gate

Particular attention should be paid to protecting the public, where necessary. Where scaffolds are erected over or next to pavements or other public places, the use of debris nets, brick guards or protective fans are essential. There should be adequate lighting and guarding of the scaffold at night.

Minimum required distances

Toe-boards should be fitted to all working platforms and are usually formed by securing a scaffold board on edge. There should not be more than 470 mm gap between the top of the toe-board and the mid guard-rail.

Working on scaffolds

Before starting work on scaffolds check that the scaffold has been inspected within the previous seven days, that the access route is safe (stair towers and ladders) and there are no gaps between boards. Do not overload loading bays or working platforms and secure loose materials.

A proprietary access system incorporating ladder access, working platform, guard-rails and hoist

Working at height

Mobile tower scaffolds and access platforms

The use of lightweight aluminium mobile towers has become a common alternative to other means of access to height on construction sites.

However, mobile towers have some limitations and should only be used when they can satisfy both legislative and general site requirements.

Any person erecting a mobile tower must be competent to do so, having received adequate training or, if not fully competent, be under the supervision of an experienced and competent person.

A risk assessment must be carried out to determine whether or not a mobile aluminium tower scaffold is suitable for the type of work that is to be carried out and the environment in which it is to be used. Factors that should be considered when deciding whether or not it is safe to use a mobile tower include are shown below.

- ☑ Is there a safer method, as far as is reasonably practicable, of carrying out the work at height?
- ☑ Will it provide sufficient height and working space? The manufacturer's instructions must be followed with regard to the maximum height to which a tower may be built, including where necessary the need to fit outriggers.
- ☑ Will it be able to take the required loading of (possibly) people, tools, equipment and materials?
- ☑ Are the ground conditions suitable (flat and firm)? Some towers have height adjustment on the legs that will, to an extent, overcome the need for a flat surface.
- ☑ Could high winds, the nature of the work being carried out or hoisting materials up from below impose forces on the tower that would try to overturn it?
- ☑ If the weather, hoisting materials or the nature of the work could cause a problem, can the tower be tied to the adjacent structure and if so, is it safe to do so?
- ☑ Are there adjacent overhead power lines?
- ☑ Is there a chance of the tower being struck by mobile plant or other vehicles?
- ☑ Is a mobile tower suitable and safe to use in all other respects?

When in use the safety factors that must be complied with include:

- ☑ statutory inspections of the tower carried out by a competent person and inspection reports compiled as necessary
- ☑ a fully boarded working platform with guard-rails and toe-boards fitted
- ☑ access to the working platform gained by using the built-in ladder sections, with users climbing the ladder on the inside of the tower (never gain access to the working platform using a freestanding ladder leaning against the tower)
- ☑ the hatch being closed as soon as everyone using the tower is on the working platform
- ☑ never gaining extra height by using a ladder, stepladder or other forms of hop up systems on the working platform
- ☑ if the job cannot be reached from the platform of the tower erected to its full height, it is the wrong tool for the job

Working at height

- [✓] the brakes being effective and set in the ON position at all times that the tower is not being moved
- [✓] the tower not being moved whilst anyone is on the working platform (equipment, tools and materials should also be removed)
- [✓] mobile towers not being moved by someone on the platform pulling the tower along using the adjacent structure.

Working platforms should be free of hazards

 For further information refer to GE 700 *Construction site safety*, Chapter D04 Scaffolding.

Podium steps

Since the coming into force of the Work at Height Regulations, greater emphasis has been placed on safe access to places of work at height, with the greater use of equipment (such as podium steps) for jobs where a stepladder or hop up would have been used previously. Podium steps offer a stable working platform, complete with wheel-brakes (but often not swivel brakes), guard-rails and in many cases stabilisers to increase the bearing footprint, in contrast to the potential instability of ladders and stepladders.

 Manufacturers emphasise that in general podium steps were introduced as a stepladder substitute and should not be regarded as anything other than this. Research published by the HSE in 2009 suggests that low level access MEWPs are safer than podiums in many situations and good planning can make their use cost effective.

Working at height

Podium in use, correctly assembled, wheels locked and displaying inspection tag

Ladders and stepladders

Traditionally, ladders and stepladders have been used as:

- ✓ the way of getting up to or down from a place of work at height (for example, the means of access to a roof or scaffold)

- ✓ a place of work at height (work carried out whilst standing on a ladder or stepladder, for example, painting first floor windows).

✓ ladders and stepladders must be inspected before use to ensure there are no defects.

Sherpamatic type steps with safe working platform and additional outriggers

The Work at Height Regulations require that employers give adequate consideration to the safety of the user before selecting a ladder or stepladder as either a means of access to height or as a place of work at height.

In deciding whether or not a ladder or stepladder should be used, the employer must carry out a risk assessment and be able to

Working at height

demonstrate that it is not reasonably practicable to use an alternative, safer means of access and that the risks from using the ladder or stepladder are low.

HSE guidance is that ladders and stepladders should only be used as a place of work when the nature of the work:

- ☑ is of short duration (a few minutes rather than hours)
- ☑ is of a light nature (requires no heavy lifting, carrying or a destabilising pressure applied by the user or equipment in carrying out the work – minimal manual handling)
- ☑ allows one hand to be available at all times for holding onto the ladder or stepladder
- ☑ requires nothing to be carried that would cause instability of the ladder, stepladder or user
- ☑ does not necessitate using the top third of the ladder or stepladder.

Additionally, when stepladders are used as a place of work they should be positioned so that the user faces the work as the stepladder is climbed. A stepladder must not be positioned so that the user is side on to the work where the nature of the work would apply a sideways pressure and cause the stepladder to become unstable.

Ladders should:

- ☑ be set up at an angle of 75° (1 m out for every 4 m up)
- ☑ be positioned on a firm, level surface
- ☑ not be vulnerable to impact by pedestrians or traffic
- ☑ be used in accordance with the manufacturer's instructions
- ☑ be the right ladder for the job

- ☑ ideally, be lashed at or near the upper point of rest by the styles, not the rungs
- ☑ extend at least 1 m above the stepping-off place unless an alternative handhold is provided that enables a safe transfer between ladder and stepping-off place
- ☑ be secured at the bottom or footed* if lashing at the top is not possible
- ☑ not be rested against fragile or flexible items (such as plastic guttering); a ladder stay or stand-off device must be used as necessary
- ☑ never be painted to an extent that the paint could conceal defects
- ☑ be subjected to a schedule of periodic inspections with written inspection reports kept, be individually identifiable and proof of inspection demonstrated
- ☑ be visually checked by the user for obvious defects before use and not used if found to be defective.

 Users should face the ladder at all times when climbing up or down, and not carry anything that would interfere with their safety or balance.

* HSE research has shown that to be most effective, the person footing a ladder should stand on the bottom rung with both feet at all times. Even then, footing is not an effective method of stopping a long ladder from slipping sideways and furthermore, two people are on the ladder, which goes against good practice. Ideally a ladder will only be footed when it is climbed for the first time for the purpose of tying it off.

Working at height

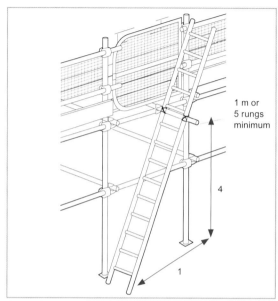

Correct ladder positioning

Trestles and lightweight staging

This type of access equipment should only be used where all of the requirements of the Work at Height Regulations can be met. Modern trestle systems are lightweight, stable in use and can be fitted with guard-rails and toe-boards as necessary.

The use of loose scaffold boards supported on split-head trestles with no means of preventing falls that could cause an injury to the user is totally unacceptable and should not be considered as an option for working at height.

Trestles should be:

- ☑ only erected by someone who has been trained to do so and is competent
- ☑ of sound construction
- ☑ only erected on a surface that is sufficiently level and will bear the weight of the trestle plus any loading of persons and materials.

Mobile elevating work platforms

Mobile elevating work platforms (MEWPs) provide a temporary working platform where ladders would be unsafe and scaffolding is not economical or reasonably practicable.

Any work involving the use of a MEWP is classified as a *lifting operation* as defined in the Lifting Operations and Lifting Equipment Regulations, so the conditions of those regulations must be complied with.

 The manufacturer's instructions must be followed.

The limits of elevation and outreach (the operating envelope) are described in manufacturers' handbooks, so always plan your work according to these guidelines.

Ground conditions are critical to safe operation. Many platforms are fitted with outriggers or stabilisers and these must always be used in accordance with the manufacturer's instructions.

When positioning a MEWP, the following safety factors must be considered.

Working at height

- ✓ Is the machine level?
- ✓ If not, can it be levelled up?
- ✓ How firm is the ground? Will it support the MEWP and its expected loading?
- ✓ Are there cellars, basements, sewers, drains, manholes, old trenches, un-compacted backfill or anything else that might collapse under the weight of the MEWP and its load?

MEWP in use

Safety harnesses must be worn by MEWP operators (and passengers) where shown to be necessary by the findings of a risk assessment. In most circumstances, those going aloft in a cherry picker would be expected to wear a safety harness. In some circumstances the risk assessment may show that the wearing of a safety harness by the users of, for example, a scissor lift is not necessary. The decision will be based upon the assessed risk of falling from the machine, the machine toppling and the implications of being clipped to the machine in the event of an accident of any kind.

When a safety harness is worn it must be attached by a short lanyard to a strong-point fitted on the machine for that purpose. The length of lanyard used will depend upon the range of movement required, whilst preventing a fall.

The free end of the harness must never be attached to a point on an adjacent structure, as this would lead to difficulties if the platform were to be inadvertently lowered with the operator still attached to the structure, or if the ground-level controls had to be used because the operator was unable to do so for any reason.

Sheet materials and cladding will act like sails in windy conditions and can affect the stability of the platform and safety of the worker in the platform.

Never use a MEWP:

- ✗ as a jack to support anything else
- ✗ for the transfer of goods
- ✗ as a crane or lifting appliance.

 Only trained operators should use the platforms.

Operators should be trained and competent on the type of machine they are required to use and:

- ✓ be issued with a company licence or authority
- ✓ be registered with the Construction Plant Competence Scheme (CPCS) or another industry-recognised scheme.

Working at height

Personnel being considered as operators should meet certain requirements. They should:

- ✓ be sufficiently fit and mobile to climb in and out of platforms
- ✓ have a good head for heights, good hearing and colour vision
- ✓ be able to judge space and distance
- ✓ have good eye/hand co-ordination and manual dexterity
- ✓ be generally in good health.

Any medical history of fainting or dizziness may make a person unsuitable for operating a machine like a MEWP. **If in doubt, contact the Employment Medical Advisory Service (EMAS)**, via their local HSE office, or request that the person seeks advice from their GP.

Maintenance and inspections should be properly scheduled.

There should be:

- ✓ a daily inspection, usually by the operator who must have been trained and competent to do so
- ✓ a weekly inspection by a maintenance fitter or trained and competent operator
- ✓ a six monthly, or 1,000 hour, examination by a competent person, who should issue a written report.

Employers are required to keep records under the Lifting Operations and Lifting Equipment Regulations.

Emergency controls

Both ground and work platform controls should be checked as part of the pre-use inspection to ensure they are working correctly. This check should also include the emergency lowering controls that are provided to enable the work platform to be safely recovered to ground level.

The operator, or their supervisor, should ensure that a responsible person who is familiar with the emergency lowering system, is always in close proximity to the MEWP to lower the work platform, in the event that rescue is required.

Even experienced operators sometimes have difficulty locating the emergency descent controls that every mobile elevating work platform is fitted with.

 The emergency descent symbol below (available to download free from www.ipaf.org) is a practical visual aid and a prime example of an industry initiative to make access equipment even safer.

Emergency descent systems can be found on all types of mobile elevating work platforms. They differ in terms of where they are located on specific machines and how they operate.

The descent symbol should be positioned to clearly indicate the location of the emergency descent controls. Operators should ensure that somebody at ground level is properly trained on how to use the controls in an emergency.

Personnel must never attempt to climb out of an elevated MEWP if the emergency back-up system fails to work, but should stay in the work platform until rescued by other means.

Working at height

Wind

Wind speeds in excess of 25 mph can create unsafe working conditions. Winds can funnel and eddy around buildings, causing turbulence, which again may make working places unsafe. In cold weather conditions, the wind will effectively lower the temperature considerably – this is known as the wind chill factor.

Beaufort wind scale for use on land (Numbers 1–9)

Windforce number	Description of wind	Wind locally	Speed mph	Speed m/sec
0	Calm	Calm, smoke rises vertically.	1	0–1
1	Light air	Direction of wind shown by smoke drift, but not by wind or weather vanes.	1–3	1–2
2	Light breeze	Wind felt on face. Leaves rustle. Wind or weather vanes move.	4–7	2–3
3	Gentle breeze	Leaves and small twigs in constant motion. Wind extends light flags.	8–12	3–5
4	Moderate breeze	Wind raises dust and loose paper. Small branches move.	13–18	5–8
5	Fresh breeze	Small trees in leaf begin to sway. Little crested wavelets form on inland waters.	19–24	8–11
6	Strong breeze	Large branches in motion. Umbrellas used with some difficulty.	25–31	11–14
7	Near gale	Whole trees in motion. Becoming difficult to walk against the wind.	32–38	14–17
8	Gale	Twigs break off trees. Progress is generally impeded.	39–46	17–21
9	Strong gale	Chimney pots, slates and tiles may be blown off.	47–54	21–24

Roof work

When working above ground level, there is no safe height.

A worker could be killed or seriously injured in a fall from less than 2 m. Prevention is better than cure – risk assessments and method statements will, if properly carried out, result in a safe system of work.

Working at height in a safe environment

In a three-year period, 63 people died in accidents during roof maintenance. None would have died if simple precautions had been taken. 30 deaths were due to falls through fragile roof materials and 12 to falls from sloping roofs. Falls from flat roofs also occurred. The cause of these accidents was the absence of a safe system of work.

Often the work is of very short duration but a safe system must always be established as the risk of a fall is the same for a five-minute job as it is for a five-day job.

Many projects will involve roof work at some stage, whether the job is new build, maintenance, refurbishment or extending an existing building or, carrying out surveys.

The main hazards include:

- ☑ falls from the edges of flat or sloping roofs
- ☑ falls through fragile roof materials
- ☑ falls from ladders and scaffolds.

Work on roofs is often carried out in situations where the hazards involved place others, including members of the public, at risk. Poor weather conditions can add to existing dangers.

Safe systems of work

Before work starts a risk assessment must be carried out to identify the hazards and enable control measures to be put in place. A safe system of working will be derived from this, which may be written down in the form of a method statement. The safe system of working must be communicated to everyone involved in the job.

If the work is to be sub-contracted, ensure that the sub-contractor understands this requirement.

Equipment

Make sure that all equipment used when working at height is safe, tested and well maintained. Make doubly sure that the equipment is right for the job and only used as intended.

The booklet *Health and safety in roof work* (HSG33) contains expert advice from the HSE and is a useful source of further information for companies that carry out roof work.

Working at height

Training
Training produces well-instructed, safe-working employees. Without proper and specific training in roof work, employees would be a liability at height, not only to themselves but to other workers.

Supervision
A roof is no place for an unsupervised trainee. Always ensure that all roof workers are adequately and properly supervised.

Flat roofs
If there is no integral safety rail, high parapet or other effective barrier, edge protection must be provided. This may take the form of:

- ☑ a working platform around the external perimeter of the roof complete with guard-rails and toe-boards, or
- ☑ securely anchored double guard-rails and toe-boards, positioned on the roof and set as far back from the edge as possible whilst allowing the work to take place.

 Always consider the wind chill factor. Hands, fingers and feet that cannot feel can cause an accident.

At times it may be necessary to temporarily remove the edge protection (for example, during the landing of materials by telehandler, or if work has to be carried out at the roof edge). If so, the safe system of work, established from the risk assessment, must specify the alternative method of fall prevention to be used whilst the edge protection is missing. The edge protection must be reinstated as soon as possible.

Where it is not practical to install edge protection of any description it may be necessary for a harness and lanyard to be worn by the person(s) at risk of falling. If so, the free end of the lanyard must be secured to a safe anchor point (such as a latch-way system or a dead weight anchor). A short restraint lanyard might be used thus preventing the wearer from approaching too close to the edge, depending upon the range of travel required by the person on the roof.

Sloping roofs
Sloping roofs are those over 30 pitch, or less if slippery conditions prevail. Work should only be carried out by people who are:

- ☑ physically capable
- ☑ appropriately trained.

When working on sloping or slippery roofs, always use:

- ☑ a roof ladder or temporary work platform that is securely fixed and a catch barrier or a platform, or
- ☑ a three-board (600 mm) wide working platform with guard-rails and toe-boards.

On steep roofs, those over 50°, a working platform at the eaves is essential, preferably with an additional third handrail.

Working at height

Roof access system

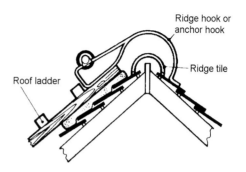

 Do not set ridge hooks onto ridge or capping tiles as the downward pressure of the ladder and a person's weight will cause the tiles to loosen.

Avoid roof work in high winds – remember that the eddying and funnelling effects of wind, which can be caused by nearby buildings and pitched roofs, can make a roof dangerous in windy conditions. The handling of sheeting and cladding at heights in windy conditions can be dangerous – for those on the roof and on the ground.

Roof ladders and temporary work platforms

It is essential that roof ladders and temporary work platforms are:

- ☑ good, well maintained and strong enough for the job
- ☑ properly supported
- ☑ securely fixed against slipping or being dislodged.

In addition, temporary work platforms should be a minimum of 600 mm wide and fitted with double guard-rails and toe-boards on both sides, unless the nature of the work requires that they are left off one side, in which case alternative fall-protection measures must be taken.

When moving across a roof, where one temporary work platform is not sufficient, use two platforms: one to work from and one to move across, ready for your next change of position.

Roof scaffolds

This type of working platform is used where work has to be carried out (for example, on a house chimney above a pitched roof). Such access platforms are usually a form of lightweight aluminium scaffold that is designed for the specific purpose. These platforms are now used as a safe means of access, where at one time an improvised working platform may have been used. Alternatively, safe access platforms conforming to the regulations may be built from tube and fitting scaffold.

This type of work has been the cause of many serious accidents because a safe means of access and/or a stable working platform with guard-rails and toe-boards was not used.

Working at height

 This type of work must never be attempted by standing on the roof and/or standing a free-standing ladder on the roof and leaning it against the chimney.

 Remember: fragile roof incidents can be prevented by careful planning, using trained and experienced workers with suitable equipment and ensuring a high level of supervision.

Fragile roofs

On average, seven people are killed each year as a result of falls through fragile roof material. Many other people suffer permanent disability. Deaths caused by falls through fragile surfaces occur mainly to those working in the building maintenance sector, while they are carrying out small, short-term maintenance and cleaning jobs. When working on roofs be aware of fragile surfaces such as:

- ☑ glass
- ☑ plastic
- ☑ asbestos-cement
- ☑ corroded metal sheets
- ☑ rotten chipboard (or similar materials)
- ☑ liner panels on built-up sheet roofing
- ☑ roof lights (which maybe difficult to see in certain light conditions)

or any other material that may fail under a person's weight. They may not be immediately obvious as they may be dirty, covered in moss or debris or they may have been painted. Reports of accidents show that skylights are a particular problem. Signs should identify all fragile roofs.

A clear indication of a fragile roof and the measures necessary to work safely upon it

Supervisors should ensure that work does not proceed on or near to a fragile surface until the appropriate measures, as highlighted by the findings of a risk assessment, have been taken. Ideally a safe system of work would be devised to enable the job to be carried out without anyone actually having to go on to the roof.

Working at height

Before work starts:

- [✓] ensure that a competent person has assessed the roof
- [✓] ensure that the work is properly planned in advance
- [✓] plan to use non-fragile assemblies for new and replacement roofs where possible
- [✓] satisfy yourself that you have allowed sufficient time to carry out the work safely.

After work starts:

- [✓] ensure the planned safe system of work is implemented
- [✓] monitor progress and maintain safe systems of work adapting, where necessary.

Where it is necessary for someone to go on the roof:

- [✓] measures must be taken to spread the loading on the roof sufficiently to prevent failure of the roof material (for example, a stable, temporary working platform with guard-rails)
- [✓] if necessary a soft landing fall-arrest system (safety nets, airbags and so on) should be positioned below the area where work will be carried out
- [✓] load-bearing covers that cannot be dislodged should be fitted over skylights, particularly where fragile skylights are fitted to an otherwise load-bearing roof

- [✓] install perimeter edge protection and use a stable, temporary working platform with guard-rails on the roof surface to spread the loads
- [✓] ensure that all the work and access platforms are fitted with guard-rails
- [✓] if this is not possible, install safety nets or air bags underneath the roof or use a harness system
- [✓] where harnesses are used, make sure they have adequate anchorage points and they are properly fitted.

Proprietary covers, which can serve as either a temporary or a permanent installation, are available to prevent someone who is passing by, or working near, fragile material from falling through.

For further information refer to:

INDG284 *Working on roofs*
www.hse.gov.uk/pubns/indg284.pdf

GEIS5 *Fragile roofs*
www.hse.gov.uk/pubns/geis5.htm

GE 700 *Construction site safety* Chapter D02
Safe working on roofs and at height

Working at height

Fall-arrest systems

Safety nets

Ideally, safety nets will be rigged immediately below where people are working to reduce the distance that anyone could fall. If this is so, anyone who falls into the net should be able to simply scramble out of it. Even rescuing someone who is injured or unconscious should not be too much of a problem.

However, if there is the potential for a fall of a significant distance into a safety net **the risk assessment must address the problem of how someone who has fallen into the net is going to be rescued, including if the person is injured or unconscious.**

When nets are used they should be carefully sited and erected by qualified persons working to BS EN 1263-2. Working platforms should still be provided if at all possible. Safety nets should be inspected weekly by a competent person.

A net that has arrested a significant fall will probably be deformed. No-one should be allowed to work over the net until it has been examined and replaced if necessary.

When working over pedestrian or traffic routes, the mesh of the net should be:

- ✓ small enough to catch falling tools, materials and debris
- ✓ strong enough to hold a falling person.

This usually involves overlaying the main net with a fine-mesh debris net.

Safety nets installed prior to roof works commencing

Soft-landing systems

Where it is not possible to devise a safe system of work that would prevent a fall from height, it is necessary to put measures in place that will:

- ✓ arrest any fall that does occur
- ✓ as far as possible limit the height that anyone could fall
- ✓ minimise any injuries resulting from a fall. Soft-landing systems are an effective way of achieving this.

A stable working platform for chimney construction

Soft-landing systems do not prevent falls but they do reduce the potential consequences of any fall. However, the benefits that these systems have, include:

Working at height

- ☑ they are passive safety systems and once installed there is nothing that the people working above them have to do to make the system effective

- ☑ they are collective safety systems that will protect everyone working above them. As such, they are much preferred to safety harnesses and lanyards that only protect the wearer, who must clip on to a safe, strong point for the system to be effective.

Air or beanbags

Designed to be used in buildings with a storey height of up to 2.5 m, these systems comprise large polypropylene bags that are located at ground floor level and either inflated with air from a pump (airbags) or are pre-packed with polystyrene chippings (beanbags). The depth of the bag cushions the fall and reduces the distance of that fall.

The bags are linked together with plastic snap-clips to completely fill the ground floor area. They can also be used on the first or subsequent floors providing that the floor joists supporting the bags are boarded over.

Safety harnesses and lanyards

If fall-prevention measures (working platforms, barriers, guard-rails and so on) or collective fall-arrest measures (safety nets or other soft-landing systems) are not practical, an alternative safe system of work must be designed. This safe system may require the use of safety harnesses and lanyards, but these should be regarded as the last resort.

Care must be taken when planning to use a safety harness, lanyard and energy-absorbing system since, depending on where the lanyard is anchored, a falling person may fall around 4 m before the fall is arrested.

One of the limitations of using such fall-arrest equipment is that it only protects a person if they wear the harness properly and connect the free end of the lanyard to an appropriate and secure point. The use of any such system requires a high degree of training, competence and, initially at least, supervision. Safety harnesses must not be used in any lone-working activity as there would be no means of rescue (*see below*).

Inspection

The material used for manufacturing harnesses and lanyards is susceptible to damage from abrasion, strong acid or alkali chemicals, excessive heat, burns and ultra-violet light (sun). Damage to the stitching and fittings (such as karabiners), are other common defects. A competent person must decide whether a damaged harness or lanyard can be repaired or should be destroyed.

 Refer to the HSE's *Inspecting fall arrest equipment made from webbing or rope* (INDG367).

An accident occurred in 2002 in which a person who was wearing a harness and lanyard, and who was clipped on, fell from a roof but died when the material of the lanyard snapped due to the shock loading imposed by the fall. The HSE investigation revealed that the lanyard was defective. Subsequent HSE research showed that a 1 mm cut in the type of webbing used to make harnesses and lanyards can reduce its strength by between 5–40%.

Full body harness with energy absorbing lanyard

The user of a harness and lanyard should inspect them visually and use the fingers to detect defects in the webbing, each time before use. It there is any doubt they must not be used.

D 20

Working at height

Detailed examinations by a competent person should be carried out at intervals of between three and six months. A record of these examinations should be kept.

Types of safety harnesses
There are many types of safety harness and the employer's risk assessment will determine the correct type of harness to be used for any particular job.

Rescue
It is vitally important that anyone suspended in a harness after a fall is rescued in the shortest possible time. A medical condition known as suspension trauma can affect any person suspended in an upright position with the legs dangling, after being suspended for a prolonged period of time. Suspension trauma, caused by pressure on the legs, will cause severe discomfort and could eventually result in death.

Manufacturers of fall-arrest equipment have developed several types of rescue system that either enable the suspended person to relieve the pressure on their legs or allow them to be quickly and safely raised back to the working platform or lowered to the ground. If immediate rescue of an unconscious suspended person is not possible, but they can be reached by rescuers, the condition of the suspended person can be greatly assisted by their legs being raised.

Professional medical help must be summoned for anyone who has been suspended in a harness, become unconscious and then been rescued.

Working over or near to water

Where there is a risk of persons falling from a structure into water, a secure form of fencing, barrier or fall-arrest equipment (preferably safety nets) must be provided. This can be briefly removed for access and movement of materials, but must be replaced as soon as possible.

Other points to consider when working over water are:

- ✓ ensuring that a risk assessment has been carried out
- ✓ where possible, a suitable working platform is provided
- ✓ safety nets, if used, must be properly erected and periodically inspected
- ✓ warning notices must be placed near to all edges
- ✓ adequate lighting must be provided as necessary
- ✓ special care must be taken in inclement weather (such as fog, frost, snow and rain)
- ✓ special attention must be paid to the possibility of tides or storm surges changing water levels or flow rates
- ✓ lifejackets must be provided, and worn where there is a foreseeable risk of falling into the water
- ✓ it is preferable that all operatives can swim
- ✓ suitable rescue equipment must be provided and maintained
- ✓ a rescue plan must be devised and periodically practised
- ✓ frequent checks must be carried out to ensure that the correct number of personnel can be accounted for
- ✓ all persons must work in pairs, or in larger groups, as necessary
- ✓ all persons must be trained in the procedures for raising alarms and in rescue drills.

Working at height

 If work is being carried out from a boom-type MEWP (cherry picker) over or near to water, the operative(s) must not be clipped to the machine via a safety harness and lanyard if there is a chance that the basket could end up in the water, with a risk of drowning, if the machine was to overturn.

Working above other people

When working at height above an area where other people have access, it is important to implement a safe system of work, to prevent people below from being struck by falling objects. This may be achieved by:

- ✓ implementing a system of work that prevents anything falling
- ✓ using a combined safety/debris net to catch falling objects
- ✓ where possible, excluding people from the area below.

 For further information refer to GE 700 *Construction site safety,* Chapter D02 Safe working on roofs and at height and Chapter D05 Fall arrest and suspension equipment.

Rescue boat available for use

Working at height

21
Excavations and buried services

What your employer should do for you	298
What you should do as a supervisor	299
Excavations	300
Safe systems of work	301
Supports	301
Inspection and examination	303
Guarding excavations	303
Personal protective equipment	304
Occupational health issues	304
Buried services	305
Working in the roadway	309

Excavations and buried services

What your employer should do for you
1. Provide suitable and sufficient risk assessments and method statements for the work, including a permit to work system as appropriate.
2. Ensure that everyone is equipped with appropriate personal protective equipment (PPE) and respiratory protective equipment (RPE).
3. Arrange for the monitoring of any changes in the soil condition surrounding the excavation.
4. Provide supports for the excavation as necessary.
5. Ensure that there is safe ladder access and egress to and from the excavation.
6. Locate buried services and provide relevant information to the work team.
7. Train the work team in safe excavation techniques to expose buried services.
8. Provide barrier material, warning signs and lights for the excavation.
9. Plan for vehicles and/or plant to approach the excavation safely.
10. Restrict vehicles, scaffolds, plant (either mobile or parked) or materials from becoming too close to the edge.
11. Plan for any possible anticipated surcharging and actions to be taken.
12. Ensure that any occupational health issues, as a result of the excavation work, are controlled.
13. Plan for statutory inspections to be carried out and inspection reports raised as necessary.

D21

Excavations and buried services

What you should do as a supervisor

Checklist	Yes	No	N/A
1. Take note of risk assessments and brief your work team to follow method statements for the work, including issuing a permit to work as appropriate.			
2. Provide everyone with appropriate PPE and RPE.			
3. Monitor any changes in the soil condition surrounding the excavation and report any areas of concern, including withdrawing workers.			
4. Ensure that supports for the excavation are used properly.			
5. Check that there is safe ladder access and egress to and from the excavation.			
6. Continuously check for buried services and take note of relevant information.			
7. Check that the work team are trained in safe excavation techniques to expose buried services.			
8. Install and maintain barrier material, warning signs and lights for the excavation.			
9. Organise for vehicles and/or plant to approach the excavation safely.			
10. Prevent vehicles, scaffolds, plant (either mobile or parked) or materials from becoming too close to the edge.			
11. Avoid any possible surcharging and pre-plan actions to be taken in the event of surcharging.			
12. Check that any occupational health issues, as a result of the excavation work, are monitored.			
13. Carry out statutory inspections and prepare inspection reports as necessary.			

D21

Excavations and buried services

Excavations

Every year people die and many others are seriously injured in collapsed trenches. A trench looks like a grave but that is where the similarity must end.

Most deaths occur in trenches less than 2.5 m deep.

An experienced, competent person must always be responsible for the management of work in excavations, including arranging for the installation of a suitable trench support system where necessary.

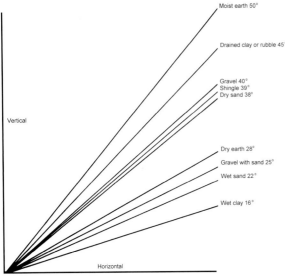

Different types of ground composition and soil, and to an extent the prevailing weather conditions, will dictate how the trench may be dug and supported.

Where possible, it is far better to shore or batter back the sides of an excavation to a safe angle, based upon the chart on this page, and therefore eliminate the chances of the sides collapsing.

 A small collapse of earth into a trench may only involve a cubic metre of soil, but that will weigh over a tonne. If the ground is heavy clay and waterlogged, it will weigh much more. In most cases, digging someone out who has been buried by a fall of material will not be a quick job. Spades and shovels cannot be used too close to the casualty. If they have been totally buried, the chances are that they will die.

The sides of a trench may look firm, but looks can be deceptive. Besides the inherent nature of the ground being excavated, additional factors that can cause what was thought to be a stable excavation to collapse include:

- ☑ heavy rain waterlogging the ground
- ☑ the ground drying out and shrinking
- ☑ excess pressure (known as surcharge) caused by nearby vehicles, materials or structures, including scaffolds.

The longer that an unsupported excavation is open, the more chance there is of a collapse.

Excavations and buried services

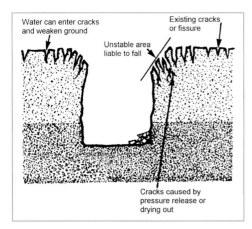

Potential dangers of trenches

Health and safety regulations no longer stipulate a depth at which the sides of a trench must be supported; the decision must be based upon the findings of a risk assessment. In some cases, shallow trenches will not need support, providing the type of ground is firm and self-supporting.

Safe systems of work

Work in an excavation has the potential to be a hazardous activity. A risk assessment must be carried out to establish the inherent hazards and how they can be overcome.

Depending upon the nature of the work to be carried out, it may be necessary to draw up a method statement and/or implement a permit to work system.

 For further information on method statements and permits to work refer to Chapter A04 Risk assessments and method statements.

In addition to the collapse of the sides, other potential hazards that may need to be addressed are:

- ☑ falls of materials (including deliberate tipping), people and plant/vehicles into the excavation
- ☑ undermining nearby structures
- ☑ accidental or deliberate contact with underground services
- ☑ in-rush of water or other fluids, standing water and pumping out
- ☑ accumulation of heavier-than-air gases (such as liquefied petroleum gas (LPG))
- ☑ seepage of naturally occurring gases (such as methane).

Supports

Choice

There are several types of trench support system. Metal support systems, with or without hydraulic adjustment, are increasingly used in preference to timber supports. The decision as to what is adequate support depends on the type of excavation and the nature of the ground. Initial risk assessments, survey and planning must be carried out by trained and competent people. Where an excavation is sufficiently deep it is a job for a specialist engineer.

Excavations and buried services

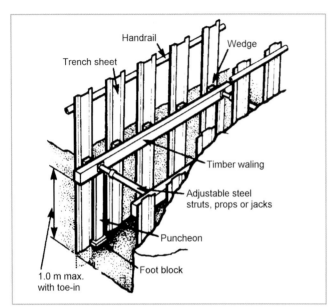

Example of a trench support system

Stability of adjacent structures

There are laws requiring that attention is given to the stability of and encroachment upon nearby buildings and, in some cases, scaffolds, not only at the planning stage but throughout the whole job. If an excavation is dug adjacent to the wall of a building, the removal of earth or undermining may cause the foundations and wall to slip or collapse.

Installation of trench supports

Placing, alteration and removal of trench supports must only be carried out by competent workers or other workers under the strict supervision of a competent person.

Safe working in a deep excavation using trench boxes

Excavations and buried services

Safe access into a deep excavation, together with all round secure edge protection

Inspection and examination

A competent person must inspect an excavation:

- ☑ before any person carries out any work in the excavation
- ☑ at the start of every shift
- ☑ after any event likely to have affected the strength or stability of the excavation
- ☑ after any accidental fall of rock, earth or other material.

An inspection must also be carried out at intervals not exceeding seven days and a report of the inspection completed as required by the Construction (Design and Management) Regulations.

Guarding excavations

Barriers

It is necessary to erect barriers around most excavations, particularly those that are in a place to which the public has access. Barriers must be erected around any excavation into which a person could fall and suffer personal injury, identified by the findings of a risk assessment. On many sites it will be common practice to install guarding around all excavations. Particular attention must be given to the guarding and even covering of excavations if there is a chance of children gaining access to the site.

Excavation edge protection barriers

303

Excavations and buried services

Excavations may be guarded by the installation of a substantial barrier, comprising double guard-rails and toe-boards or fencing on all sides. Alternatively, a barrier may be formed by depositing the excavated spoil in an extended heap along the length and width of the excavation to prevent vehicles from getting too close.

Lighting

During the hours of darkness or other periods of reduced natural light, excavations or the guarding surrounding them may need to be lit, particularly if the excavation is in a place to which the public has access.

Access

There must be adequate and safe arrangements for getting into and out of excavations. This is usually with ladders, properly located and fixed. Under no circumstances should anyone use the side supports or underground services that cross or run along the trench as footholds to climb into or out of an excavation.

Surcharging

The layout of the site must be planned to avoid, as far as possible, the surcharging (applying excessive pressure) to the sides of any excavation. This might involve planning vehicle routes, parking areas and material lay-down areas to keep them well away from where excavations are to be dug. Where surcharging is inevitable, the way in which the sides of the excavation are supported must be designed and installed to accept the load.

Vehicle and plant movements

Where it is necessary for plant or vehicles to approach an excavation (for example, to tip materials or actually dig the excavation), adequate measures must be taken to ensure that the activity can be carried out safely.

Examples of safe practice are:

- ☑ the installation of anchored stop-blocks to prevent the vehicle wheels over-running and getting too close to the edge
- ☑ using a signaller to give directions to the vehicle driver/plant operator
- ☑ ensuring that vehicle drivers/plant operators are competent.

Personal protective equipment

Depending upon the nature of the work carried out it may be necessary to use items of personal protective equipment (PPE) beyond that necessary for normal site work. The need for PPE, particularly specialist PPE, will be indicated by the findings of the risk assessment for the job.

It may be necessary to wear respiratory protective equipment (RPE) in excavations where the air quality cannot be guaranteed as safe to breathe, particularly where, because of its nature, the excavation becomes a confined space.

For further information on PPE and RPE refer to Chapter B09 Personal protective equipment.

Occupational health issues

Leptospirosis (Weil's disease)

Leptospirosis is a potentially fatal disease that is caught from contact with rat urine or water contaminated by rat urine. Rats will be attracted to wet places (such as the bottom of trenches) and they are likely to be a problem if sewers or other underground pipes, where they may live, have to be broken into as part of the work.

Excavations and buried services

For further information refer to Chapter B08 Health and welfare.

Asphyxiation/poisoning

An excavation can become a confined space due to a combination of the:

- ☑ nature of the excavation (deep and narrow) and the ground in which it is dug that could result in the accumulation of naturally occurring gas (such as methane)
- ☑ type of work carried out
- ☑ substances used (such as heaver-than-air gases, for example, LPG)
- ☑ ingress of fumes from plant exhausts
- ☑ nature of the substances released from any buried services that have to be broken into.

All of these factors can potentially have the effect of reducing the percentage of oxygen in the air and/or introducing poisonous gases into the air. In many cases, occupational health problems will only be avoided by wearing the appropriate type of RPE or installing forced fresh-air ventilation.

For further information refer to Chapter D22 Confined space working.

Buried services

Plan – locate – then dig

Every buried service is a danger to the unwary. It is a tragic fact that there are many injuries and deaths each year caused by accidental contact with underground services.

The services you are most likely to find and their relevant colours are (this list is not exhaustive):

- ☑ **Electricity (all voltages)**   Black or red
- ☑ **Water**   Blue, black or grey
- ☑ **Gas**   Yellow
- ☑ **Communications**   Grey, white, green, purple or black.

Buried services

Excavations and buried services

Before digging
Always plan the operation and check with the owners of the services to see if they have underground services in the area in which you plan to dig. They may be planning to open the ground around the same time and it may be possible to carry out all of the work during a single opening.

The local electricity or gas company will give advice by telephone.

 British Telecom operates a similar type of service and those wishing to make such enquiries should dial 0800 023 2023 and select the appropriate option.

Types of marker posts commonly found to indicate the location of buried services

Plans do not always indicate the exact location of a service pipe or cable, and on occasions, services are moved without authority or consent so it pays to look out for indicator posts, manholes, valve covers and so on as clues to the route of buried services in the area in which you plan to work.

Brightly coloured plastic tape or mesh, sometimes with a metal insert, may be left about 300 mm above a buried service in the backfill. The absence of such a marker does not indicate that there is no buried service, so be careful when you are digging.

Cable and pipe locators
There is a wide range of instruments designed to find buried services, which can detect cables, pipes and metal objects.

They appear easy to use but it is essential that operatives are properly trained and use the correct instrument for the job.

A cable locator

It will not be possible to detect non-metallic objects using conventional detectors unless an end of a pipe can be exposed and a sonde (type of transmitter) used in conjunction with a generator to trace the route. Even the most skilled operator using the best equipment will not find every pipe or cable every time. In unskilled hands, the instrument may not find anything.

 Conventional cable locators will not find plastic pipes.

Excavations and buried services

The development of hand-held ground-probing radar, which has the capability to locate changes in density of the material below the transmitter, has enabled trained operators to detect non-metallic objects (such as plastic or earthenware pipes and even air-filled voids).

Digging

Depending upon the potential hazards, it may be necessary to use a permit to work system before commencing any digging. On some sites a specific permit to dig system is used.

Before starting the main dig, with or without the use of excavators, always dig trial holes by hand to establish where a buried service is located.

Never assume that a pipe or cable will run in a straight line between any two holes.

Do not use power tools or excavators within 500 mm of the indicated line of any pipe or cable. If a power tool is used to break paved surfaces, take special care to avoid striking any pipe, or similar, that may have been buried at a shallow depth.

Use spades and shovels when hand digging in preference to picks or forks, which are more likely to pierce cables and pipes.

Unless you have proof that a buried service has been made safe, assume that it is live.

Checklist

- ☑ Always check with service providers or landowners.
- ☑ Obtain a permit to dig if such a system is in use.
- ☑ Before starting, establish what type(s) of service are there.
- ☑ Always assume that services are there and that they are live.
- ☑ Use detectors and look for signs of marker tapes.
- ☑ Where possible, ensure that services have been isolated or disconnected before starting work.
- ☑ Be prepared for services that are either not where you expect them to be or not at the expected depth.
- ☑ Be wary of displaced tapes, tiles or slabs. Use spades and shovels rather than forks and pickaxes, and carefully lever out rocks, stones or boulders.
- ☑ Do not over-penetrate with power tools.
- ☑ Keep clear of digger buckets, especially when near buried services.
- ☑ Services enclosed in concrete must be switched off if the work involves breaking them out.
- ☑ Where possible the final exposure of buried services should be carried out in a way that prevents any damage (such as using a compressed air lance).

Never assume that a service is dead – always treat it as live until confirmed otherwise.

Damage to buried services

Damage to buried services **must** be reported to the owner/occupier immediately. When such damage causes an emergency situation, call the **police – fire – ambulance** services as necessary.

Gas

For gas emergencies, phone Gas Emergency Services on 0800 111999.

Excavations and buried services

Contact the police, fire brigade and gas supply company immediately. If there is a dangerous situation and the emergency services have not arrived, try to evacuate the immediate area, including if necessary, the occupants of nearby properties to an upwind position. As far as possible, prevent anyone from smoking and keep traffic clear of the area.

 If the gas escape catches fire, do not attempt to extinguish the flames.

Depth of cover

The normal minimum depth of cover for gas mains operating in the low and medium pressure ranges is:

- ☑ 600 mm in footways or verges
- ☑ 750 mm in carriageways.

These figures may vary since each gas company can have its own standards.

Electricity

 For electricity emergencies, phone the National Grid 24-hour electrical emergency service on 0800 40 40 90.

Avoid contact with any damaged cable or apparatus. If you are operating a machine do not attempt to disentangle any equipment. If at all possible, jump clear of the machine, ensuring that you do not make contact with the vehicle and ground at the same time.

If this is not possible, stay exactly where you are. As far as possible, do not touch any metallic part of the vehicle and shout for help. If you are able, inform the electricity company or ask someone else to do so. Keep people away.

Depth of cover

The depth at which electricity cables or ducts are usually installed in the ground is decided by the need to avoid undue interference or damage.

Depending on the type of cable and the power that it may be carrying, the depth of cover may vary from 450–900 mm.

However existing services can be found at any depth. It is not uncommon for them to have as little as 50 mm cover.

In all cases where the depth of cover is likely to increase or decrease, the service owner must be consulted.

Excavations and buried services

Other services
Leave a damaged service well alone and inform the owner.

Backfilling
Never tip hardcore or rock onto a buried service in an attempt to fill the hole more quickly. Use selected backfill, settle and compact carefully but avoid any mechanical or other shock to the pipe or cable. Place warning tapes or tiles approximately 300 mm above the service.

In the case of gas pipes, seek advice from the gas company on backfill methods.

Working in the roadway
Unless you are a licensed utility company, you may not work in the footway, carriageway or any verge. If you are working in the roadway, you must comply with the New Roads and Street Works Act.

There are two main Acts that govern works on the highway: the New Roads and Street Works Act and the Highways Act.

New Roads and Street Works Act
The New Roads and Street Works Act applies to works carried out in a street by an undertaker exercising a statutory right to inspect, place and maintain pipes, cables, sewers or drains, which are laid in the carriageway or footway. The term **undertaker** also covers holders of street works licences.

When a Local Authority acts as an agent to an undertaker or contractor to carry out work for an undertaker, the execution of the works is governed by the Act.

Highways Act
The Highways Act applies to all work on the highway. The Highway Authority (or Roads Authority in Scotland) must be consulted and grant permission for works to be carried out. This applies to any works for road construction or maintenance purposes. These are covered by the provisions of the Highways Act and are the responsibility of the Highways Authority.

This Act also makes provision for licences for skips and scaffolds and places responsibility for safety with the Highway Authority. Contractors must obtain permission before working on the highway.

Plan – the work to be done.
Locate – the services before digging.
Dig – use a safe method of work.

For further information refer to GE 700 *Construction site safety*, Chapter D07 Excavations, Chapter D08 Underground and overhead services and Chapter F01 Street works and road works.

Excavations and buried services

D 21

22
Confined space working

What your employer should do for you	312
What you should do as a supervisor	313
Legislation	314
Confined spaces	314
Hazards in confined spaces	315
Hostile environments	316
Information, instruction and training	316
Safe working	317
Selection of workers	319
Permit to work	320
Case studies	321

Confined space working

	What your employer should do for you
1.	Ensure that managers, supervisors and employees are aware that confined space working should be avoided, if possible, and if not avoidable, planned before entry.
2.	Train staff (in line with the requirements of the regulations) to be aware of the different hazards of confined spaces.
3.	Put in place a rescue plan prepared and practised prior to entry.
4.	Issue the correct personal protective equipment (PPE), respiratory protective equipment (RPE) and rescue equipment.
5.	Ensure that the work team who are to enter confined spaces are trained, competent and physically fit.
6.	Enforce a permit to work system with any necessary lock-off systems.

Confined space working

What you should do as a supervisor

Checklist	Yes	No	N/A
1. Be aware that confined space working should be avoided, if possible, and if not avoidable, planned before entry.			
2. Ensure that the work team are trained (in line with the requirements of the regulations) and aware of the different hazards of confined spaces.			
3. Supervise the work to ensure that a rescue plan has been prepared and practised prior to entry.			
4. Ensure that the correct PPE, RPE and rescue equipment is available and issued.			
5. Check that the work team who are to enter confined spaces are trained, competent and physically fit.			
6. Issue a permit to work with any necessary lock-off systems put in place.			

D22

Confined space working

Legislation

 A confined space has two defining features:

- ☑ a space which is substantially, though not always entirely, enclosed
- ☑ a place where there is a reasonably foreseeable risk of serious injury from hazardous substances or conditions within the space or nearby.

The key duty under Regulation 4 is a duty on employers to plan work that avoids the need for anyone to enter a confined space, by carrying out the work by alternative means, where it is reasonably practicable to do so.

Confined spaces

Examples of confined spaces that may be encountered during construction activities are:

- ☑ a cellar or an inadequately ventilated basement room
- ☑ a boiler, boiler flue or chimney
- ☑ a manhole, sewer, drain or excavation
- ☑ a ceiling void or duct
- ☑ a caisson or cofferdam
- ☑ an excavation
- ☑ a loft space

- ☑ any room or enclosed space with poor ventilation can become a confined space (for example, painting a room or power floating in a large hanger using petrol-powered equipment).

There are many more examples of confined spaces and the hazards will vary with the location, the type of work being carried out, and the equipment or substances used.

Lofts, enclosed/unventilated rooms or similar areas can be confined spaces

Confined space working

Hazards in confined spaces

Oxygen deprivation and suffocation

The air that we breathe contains around 21% oxygen and, at that figure, people can work without difficulty. If the oxygen level falls to 17%, ill effects start to be felt, including the loss of co-ordination and concentration, together with abnormal fatigue. Should the oxygen level fall further to 10%, it will cause breathing difficulties and unconsciousness. If it falls to 6%, death can follow quite quickly.

Oxygen deprivation may be the result of:

- ☑ the displacement of oxygen by gas leaking in from elsewhere, or the deliberate introduction of purge gas
- ☑ the displacement of oxygen by a naturally occurring gas (such as methane)
- ☑ oxidisation, rusting or bacterial growth using up the oxygen in air
- ☑ oxygen being consumed by people breathing
- ☑ any process of combustion (such as welding and other hot works)
- ☑ the prior discharge of a fire-extinguishing system containing halon or carbon dioxide.

Toxic atmospheres

However much oxygen is present in the atmosphere, if there is also a toxic gas present in sufficient quantity it will create a hazard. Some of the many toxic gases that may be encountered include:

- ☑ hydrogen sulphide, usually from sewage or decaying vegetation
- ☑ carbon monoxide from internal combustion engines, or any incomplete combustion, especially of LPG
- ☑ carbon dioxide from any fermentation or naturally evolved in soil and rocks, or coming from the combustion of LPG
- ☑ fumes and vapours from chemicals (such as ammonia, chlorine, sodium, and from petrol and solvents).

Whenever a toxic gas (or any gas, fume or vapour and so on that may be hazardous to health) is thought to be (or known to be) present, then an assessment of the risk to health must be made under the provisions of the **Control of Substances Hazardous to Health Regulations** and the appropriate control measures must be put into place to eliminate or control the risk.

 For further information refer to GE 700 *Construction site safety,* Chapter B07 The Control of Substances Hazardous to Health.

Petrol and diesel engines create carbon monoxide, which is an extremely toxic gas hazard. Liquid petroleum gas-powered engines create an excess of carbon dioxide, which is a suffocating hazard. The use of any form of internal combustion engine within a confined space must be prohibited, unless a specifically dedicated exhaust extraction system is operative.

Confined space working

Test the atmosphere before and during confined space entry

Flammable atmospheres

Some gases need only be present in very small quantities to create a hazard. A few of the major sources of explosive and flammable hazards are shown below.

- ☑ Petrol or liquefied petroleum gas, propane, butane and acetylene. These are explosive in the range of 2% in air upwards. The hazard is normally created by a spillage or leakage.

- ☑ Methane and hydrogen sulphide, which are naturally evolved from sewage or decaying organic matter. These are explosive in the range of 4% in air upwards.

- ☑ Solvents, acetone, toluene, white spirit, alcohol, benzene, thinners and so on. These are explosive in the range of 2% in air upwards. The hazard generally results from a work process and/or spillage.

- ☑ Hydrogen and other gases evolved from processes (such as battery charging).

- ☑ Airborne dust clouds.

In an explosive or flammable atmosphere, a toxic or suffocating hazard may also exist.

Hostile environments

Apart from the hazards dealt with above, other dangers may arise from the use of electrical and mechanical equipment.

Extremes of excess heat and cold can have adverse effects and may be intensified in a confined space. Further dangers exist in the sheer difficulty of getting into, or out of, and working in a confined space. The potential hazard of an inrush of water, gas, sludge and so on, due to a failure of walls or barriers, or leakage from valves, flanges or blanks, must all be considered at the risk assessment stage.

Information, instruction and training

The need for comprehensive training prior to any involvement in confined space working cannot be stressed too highly. The number of deaths in confined spaces in recent years, many of them resulting from an attempted improvised rescue of the original casualty, is testimony to the extreme hazards that can be present in confined spaces.

Confined space working

 A person should never enter a confined space unless they are trained and are competent to do so.

Not only should operatives who have to enter confined spaces be trained, but also anyone involved in:

- ☑ planning and supervising confined space work
- ☑ communication between those inside and outside the confined space
- ☑ rescue activities.

All employees who are required to work in confined spaces must be trained to carry out the work safely. The content of the training required will vary according to circumstances and the type of space being entered (for example, in some circumstances training in the use of atmospheric monitoring equipment or full breathing apparatus may be required).

The training of anyone who might have to enter a confined space (for example, workers, supervisors or rescuers) must include a significant amount of practical work in addition to the theory.

Safe working

Risk assessments

An early decision must be made as to whether it is really necessary for someone to enter the confined space or whether an alternative method for carrying out the work is practical.

Is confined space working really necessary?

A comprehensively planned approach to the management of risk in confined spaces is essential. The findings of a risk assessment will determine the safe system of work that in most cases will be:

- ☑ supported by a method statement or another form of documented procedure
- ☑ controlled, and in many cases limited in scope, by a permit to work.

Confined space working

The risk assessment, from which the safe system of work will be developed, must consider the need for:

- ☑ an assessment of the possible hazards present
- ☑ atmospheric monitoring both before entry and continuously whilst work is carried out
- ☑ appropriate respiratory protective equipment (RPE)
- ☑ a trained rescue team and appropriate rescue plan
- ☑ a means of communication between those in the confined space and those outside
- ☑ appropriately trained staff
- ☑ pre-entry notification of the emergency services.*

* HSE stress that people who plan work in confined spaces must not regard the pre-notification of the emergency services as a substitute for implementing adequate emergency arrangements.

For further information on respiratory protective equipment refer to Chapter B09 Personal protective equipment.

Employers have a duty under Regulation 3 of the **Management of Health and Safety at Work Regulations** to carry out a suitable and sufficient assessment of the risks to the health and safety of anyone who might be affected by their work activities.

RPE in use

For further information on the principles and practical aspects of carrying out risk assessments refer to Chapter A04 Risk assessments and method statements.

Where a risk assessment is being prepared for work in a confined space in which there will, or might be, toxic gases or other hazardous substances, the risk assessment must satisfy the requirements of:

- ☑ the Management of Health and Safety at Work Regulations with regard to the general principles of risk assessment
- ☑ the Control of Substances Hazardous to Health Regulations with regard to the specific threat to health from harmful substances.

Confined space working

Selection of workers

Not every worker is suitable for working in a confined space:

- ☑ facial hair can reduce the effectiveness of respiratory protection considerably, by not allowing an effective airtight seal between the breathing apparatus mask and the face
- ☑ the need to wear spectacles also has the potential to be a problem when combined with the need to wear RPE as, in the majority of cases, they cannot be accommodated within the face mask.

Rescue

The extent of arrangements necessary for emergency rescue will depend upon:

- ☑ the risks identified
- ☑ the physical nature of the confined space
- ☑ how the rescue will be carried out.

Rescue equipment will often include a lifeline, a tripod and hoist (over the entry point), breathing apparatus and first-aid equipment. Anyone who has to be rescued may well be suffering from respiratory difficulties, so the possible need for resuscitation should be considered (equipment and trained staff).

Enter only after a risk assessment

 For further information refer to GE 700 *Construction site safety*, Chapter D09 Confined spaces.

Confined space working

Permit to work

The use of a permit to work system may be appropriate for some types of work carried out in excavations (for example, where an excavation has become a confined space or where services have to be isolated before they can be disturbed or disconnected).

The permit will certify that all the necessary control measures have been taken to enable the work to be carried out safely.

See below for a possible layout for a permit to work certificate.

PLANT DETAILS (Location, identifying number, etc.)			ACCEPTANCE OF CERTIFICATE	I have read and understood this certificate and will undertake to work in accordance with the conditions in it
WORK TO BE DONE				
WITHDRAWAL FROM SERVICE	The above plant has been removed from service and persons under my supervision have been informed			Signed Date Time
	Signed Date Time		COMPLETION OF WORK	The work has been completed and all persons under my supervision materials and equipment withdrawn
ISOLATION	The above plant has been isolated from all sources of ingress of dangerous fumes, etc.			
	Signed			Signed Date Time
	The above plant has been isolated from all sources of electrical and mechanical power		REQUEST FOR EXTENSION	The work has not been completed and permission to continue is requested
	Signed			
	The above plant has been isolated from all sources of heat			
	Signed Date			Signed Date Time
CLEANING AND PURGING	The above plant has been freed of dangerous materials		EXTENSION	I have re-examined the plant detailed above and confirm that the certificate may be extended to expire at
	Material(s): Method(s):			
				Further precautions:
	Signed Date Time			
TESTING	Contaminants tested	Results		Signed Date Time
			THE PERMIT TO WORK IS NOW CANCELLED. A NEW PERMIT WILL BE REQUIRED IF WORK IS TO CONTINUE	
	Signed Date Time			
I CERTIFY THAT I HAVE PERSONALLY EXAMINED THE PLANT DETAILED ABOVE AND SATISFIED MYSELF THAT THE ABOVE PARTICULARS ARE CORRECT *(1) THE PLANT IS SAFE FOR ENTRY WITHOUT BREATHING APPARATUS				Signed Date Time
(2) BREATHING APPARATUS MUST BE WORN Other precautions necessary: Time of expiry of certificate: * Delete (1) or (2)			RETURN TO SERVICE	I accept the above plant back into service
Signed Date Time				Signed Date Time

Case studies

 Confined spaces can be killers. At Carsington Reservoir in Derbyshire, four young men all aged between 20 and 30, and physically fit, died at the bottom of an open-topped inspection shaft. Naturally-evolved carbon dioxide had displaced the oxygen, but no tests were made before the first man entered. The first man down collapsed and the three other men in turn climbed down to their deaths in futile attempts to rescue their colleagues.

 When an engineer collapsed in a sewer, a rescuer entered without breathing apparatus and was overcome. Another person made a similarly vain attempt to reach the victims. When the fire brigade arrived, they had to remove the two would-be rescuers before they could get to the engineer. By then it was too late and he died. Had everyone waited for the fire and rescue service, the engineer might have lived and the rescuers would not have needlessly died.

Confined space working

D 22

23

Environmental management

What your employer should do for you	324
What you should do as a supervisor	325
Definition of the environment	326
What is environmental management?	326
Activities that cause environmental damage	326
Managing the project environmental issues	327
Environmental impact assessments	328
Pollution prevention	328
Managing site discharges and drainage	330
Fuel and chemical storage and management	331
Dealing with environmental incidents	332
Managing transport, materials, water and energy use	333
Plants and wildlife	337
Archaeology and heritage	340
Land use and contamination	341

Environmental management

	What your employer should do for you
1.	Identify the environmental risks and control measures for the project and put in place an environmental management plan.
2.	Ensure that staff are competent to carry out their environmental responsibilities.
3.	Ensure that everyone is aware of their environmental responsibilities and put in place the correct communication channels.
4.	Ensure that site inductions and briefings include the key project environmental issues.
5.	Make sure that all environmental licences, discharge consents and authorisations are in place with the environmental regulators.
6.	Ensure sub-contractors and suppliers have considered appropriate environmental controls in their method statements.
7.	Provide suitable and sufficient spill-kit facilities and emergency reporting arrangements.
8.	Ensure that suitable protection arrangements are put in place for sensitive areas.
9.	Ensure that the project environmental performance is regularly discussed, reported and communicated.
10.	Deal with any environmental problems reported to them.
11.	Lead by example in environmental best practice.

Environmental management

What you should do as a supervisor

Checklist	Yes	No	N/A
1. Ensure you are aware of your environmental responsibilities for the area under your supervision.			
2. Ensure your manager has informed you of the arrangements for:			
• oil storage and refuelling			
• emergency sill procedures			
• noise and vibration (working hours, screening and so on)			
• water discharges			
• protection of sensitive areas (trees, vegetation, wildlife and so on)			
• dust controls (wetting down and road sweeping)			
• waste disposal (skips, waste carriers and so on).			
3. Ensure your workforce understand the environmental risks and the arrangements to control them.			
4. Carry out regular inspections of the works.			
5. Report any problems to your manager.			

E23

Environmental management

Definition of the environment

The environment can be defined as any physical surroundings consisting of air, water, land, natural resources, flora, fauna, humans and their interrelation.

What is environmental management?

Environmental management is about protecting the environment; minimising and controlling the impacts that may arise and the damage that could be caused. Building and construction site activities are likely to have some impact on the environment.

A company's environmental performance can significantly affect public perception that, in turn, may influence clients in their choice of contractors. Many well-managed companies will already have an established environmental policy in place, alongside their health and safety policy.

Benefits of environmental management

The main drive of current environmental legislation is to prevent damage, harm or pollution to living organisms or their habitats, or to human life, people's health, senses or property that can arise from the release of any substances or emissions. There are several key benefits to environmental management.

- ☑ **Meets legal requirements.** There is extensive legislation covering the environment, some of which is formed in line with the 'polluter pays' principle. Failure to comply may result in companies or individuals being fined or, in the worst case, imprisoned for causing damage to the environment.

- ☑ **Makes business sense.** As clients become more aware of environmental management, it is often expected as standard; it will enhance the company's reputation and give a better public image.

- ☑ **Saves money.** Good site management, including waste management, will reduce the cost of landfill tax and other associated costs.

Before development starts, designers and contractors should, together with the client, establish the likely environmental impacts that the project or work activities may have. They should make provisions to eliminate, as far as practicable, any sources of environmental damage or pollution, and establish sufficient control measures to minimise the harmful effects of construction activities.

There are significant opportunities at the design stage to reduce environmental impacts (for example, specifying materials with recycled content or manufacturing components off site for simple assembly on site without rework and waste).

Activities that cause environmental damage

The construction industry can damage the environment in a number of ways. It therefore has a major role to play in protecting natural resources and ensuring that they are passed on to the next generations, in good order, for their enjoyment. Construction activities with the potential to cause environmental damage are:

- ☑ environmentally-damaging designs or poor choice of materials
- ☑ high energy usage (and consequent greenhouse gas emissions)
- ☑ ill-conceived developments
- ☑ the construction process itself.

Clients, their designers and specifiers can eliminate many of these damaging activities by taking the problems into account when a project is first considered. Contractors and developers can also play their part in reducing the impact of their work, and it is important

Environmental management

that environmental matters are incorporated within company risk assessment and management processes.

Damage to the environment may arise from construction site activities, which may include, but are not limited to, any of the following.

Atmosphere	Land	Water
- Asbestos	- Asbestos	- Silt
- Dust	- Chemicals	- Chemicals
- Radiation	- Lead	- Contaminated water
- Exhaust emissions	- Litter	- Run-off
- Gases or vapours	- Oils and fuels	- Effluent
- Noise	- Spillage of materials	- Oils and fuels
- Smoke	- Waste materials	- Hazardous solid matter
		- Slurry

It is important that one solution to environmental pollution does not divert the problem to another medium (for example, a solution to air pollution must not lead to water contamination).

Managing the project environmental issues

The project manager will have overall responsibility for ensuring that the right environmental controls are put in place with input from the client, designers, suppliers, sub-contractors and other site personnel. There are five key steps to ensure that environmental issues are effectively managed on site.

Step 1. Identify the project environmental requirements. The client specification, together with other documentation (such as the planning conditions for the project) will define what environmental requirements need to be met.

Step 2. Identify the project environmental risks. Having identified the project environmental requirements in Step 1, the next step is to determine how they can be overcome on site. Emergency situations should also be considered. This step should define the key environmental risks for the project.

Step 3. Define responsibilities for managing the environmental risks. Having identified all of the project environmental risks in Step 2, the project manager will define what responsibilities and lines of communication are needed to manage the environmental risks effectively, including who will produce various environmental plans in Step 4 below.

Step 4. Put in place the controls for managing the environmental issues. Various documents and plans will draw together the project environmental requirements, risks and controls and responsibilities for managing them. This is usually in the form of a project environmental management plan (EMP). Other documents will include the site waste management plan, water management plan, noise management plan, and method statements to control specific tasks as appropriate.

Step 5. Put in place a monitoring and inspection regime. A monitoring system will be put in place to ensure that the requirements of the EMP and specific control measures are being managed effectively. Site supervision will be a key part of this process and will be involved in checking and inspecting site controls for:

- oil, fuel and chemical storage
- spill and emergency response equipment
- waste facilities and housekeeping

Environmental management

- water discharges
- dust management
- noise management
- protection of sensitive areas.

Any problems should be reported to your site manager so that they can be rectified.

Environmental impact assessments

Environmental impact assessments are a formal, systematic process used to assess the effects that a proposed development may have on the environment.

The process of environmental assessment is used to collate information for the use of the developer and planning authority in deciding if the development should go ahead. Projects that always require environmental impact assessments include:

- ☑ construction of motorways, express roads, lines for long distance railway traffic and some airports
- ☑ some waste-disposal installations
- ☑ long distance pipelines
- ☑ certain factories and manufacturing plants
- ☑ large quarries and open-cast mines.

Other projects are subject to environmental impact assessments if there is significant effect on the environment. These include:

- ☑ extraction of some types of mineral
- ☑ construction of roads, harbours, airfields, dams or other water storage structures
- ☑ industrial estate developments
- ☑ urban development projects
- ☑ yacht marinas
- ☑ urban developments, including shopping centres and car parks, sports stadiums, leisure centres and multiplex cinemas
- ☑ tramways
- ☑ coastal works (dykes, jetties, moles or other sea defence work)
- ☑ motorway service areas.

While some projects are required to have a formal environmental impact assessment completed, many others have an environmental statement to protect measures that are to be put into place during construction.

 The project manager should review the environmental statement, where available, before commencing any works.

Pollution prevention

Introduction
There are several hundred major pollution events from construction and demolition sites every year, many leading to prosecution. Common causes of pollution are poor work practices, accidents and incidents, illegal discharge, surface water run-off (rain), fire-fighting water discharge and vandalism.

Environmental management

Common substances that impact the environment include:

- ☑ concrete and cement
- ☑ grout
- ☑ oils and diesel
- ☑ silt
- ☑ sewage
- ☑ dust and smoke.

Good planning is the key to minimising the risk of a pollution event by considering whether an activity can impact on one or more of the three key receptors: **air, water or land**.

Following three simple steps, measures can be put into place to allow you to control and manage the issues.

Step 1. Identify the activities likely to cause pollution as well as the pathways and receptors on which it can impact (for example, diesel storage, surface water drains and local stream).

Step 2. Plan and put in place controls for any likely event, documenting the actions taken. Make someone responsible for the area and conduct regular inspections to make sure everything is alright (for example, locate a diesel tank in a bund away from any drains and make sure it is kept locked and there is a key holder responsible for it).

Step 3. Communicate the plans to the workforce. The site team should know what mitigation measures are in place, why they are there and what to do in an emergency (for example, for a diesel spill, locate the spill kit and contain; report and clean up the spill).

In planning for pollution control you should take account of your site neighbours, local residents, schools or other businesses.

Concrete wash water needs to be controlled

Managing noise and vibration, dust and air pollution

Nuisance is often a result of poor site practices and can lead to poor neighbourhood relations or even suspension of the works by the Local Authority. Nuisance through noise, dust, vibration, light, smoke or vermin can have a major impact. The creation of nuisances may be minimised by:

- ☑ ensuring powerful sources of lighting are directed away from residents or other sensitive receptors
- ☑ damping down site roads and regularly sweeping local roads to prevent the spread of dust; enforce site speed limits
- ☑ considering the use of wheel-washing equipment at the site exits
- ☑ minimising dusty operations by damping down or enclosing the operations within sheeting
- ☑ locating crushing plant away from sensitive areas

Environmental management

- ☑ forming holes in slabs when casting rather than cutting them out later
- ☑ ensuring noisy operations are carried out during normal working hours
- ☑ substituting noisy operations, noisy equipment or plant with quieter options or fabricate off site if possible
- ☑ ensuring equipment is properly maintained to eliminate loose parts, fitting mufflers or fixing attenuation materials to panels
- ☑ providing acoustic enclosures or screening
- ☑ locating equipment as far away as possible from sensitive receptors
- ☑ using an auger instead of driven piling
- ☑ using electric motors instead of internal combustion engines
- ☑ avoiding the use of reversing sirens or bleepers through one way haul routes if possible
- ☑ shutting down plant and equipment when not required.

Traffic and deliveries can cause congestion, noise and disruption on local roads and should be programmed to avoid queuing outside of the site. The Local Authority might request a Section 61 agreement (prior consent) that sets out measures to control noise nuisance from your site. If the conditions of the Section 61 Consent (or local limits) are breached then the Local Authority can issue a Section 60 Abatement Notice.

Managing site discharges and drainage

Where there is a possibility that effluent from any construction work might pollute drainage systems, watercourses or rivers and streams, an application for a consent to discharge must be made to the relevant authority. The issuing authority will depend on where the discharge is made.

For trade effluent you will need to obtain a trade effluent consent to discharge, obtained from the water or sewerage company or authority.

If you are going to discharge to surface water (rivers, streams, estuaries, lakes and so on) or to groundwater, you may need an environmental permit. Small quantities of some liquid wastes may be discharged under an exemption or general binding rules. Contact your environmental regulator before commencing works to seek clarification.

Consent from the environmental regulator will also be required for any works on, above, below or adjacent to a watercourse, generally within 10 m.

Have an up to date and accurate site drainage plan available to identify the location of all drains and sewers in and around your site and where they lead. When you make a discharge to a drain always check that you are connecting to the correct system:

- ☑ to the foul sewer for trade effluent (trade effluent is any liquid waste (effluent) that is discharged from premises being used for business, trade or industry) and sewage
- ☑ to the surface water drainage system for clean uncontaminated surface water.

It will help to colour code manhole covers, gullies and grills by painting them with a recognised system: blue for surface water drains and red for foul water drains.

Environmental management

You should understand where all the drains are on your site and where they lead to.

Excavations need to be considered in terms of the need to dewater and discharge. Silty water will need to be treated in settlement tanks, lagoons, filtration systems or using flocculent prior to discharge. The size of the tank/lagoon should be adequate for the settlement time required and the rate at which the water is pumped into it.

Muddy water should not be discharged into surface water drains

Concrete washout areas should be located away from watercourses and other sensitive areas and at least 10 m away from drains and watercourses. Water from concrete washout is highly alkaline and should be disposed of to foul water systems or disposed of cost-effectively elsewhere.

A robust monitoring system should be put in place to regularly check that controls are in place and required discharge quantities and contaminant thresholds are not exceeded.

Fuel and chemical storage and management

Oil and diesel spills can have a major impact on surface waters and drainage systems. There are regulations in place requiring certain types of oil storage to be in secondary containers (bunds). Bunds should be capable of holding 110% of the capacity of the tank that sits within it.

 Store all fuels and liquids in bunded areas or on drip trays, away from drains, and have spill-kit clean up materials (such as absorbent pads and booms) on site.

COSHH stores are used on some sites, where hazardous chemicals and substances are safely stored and environmental risks minimised.

Solvents, paints and chemicals should be stored in accordance with their COSHH/material safety datasheets. The floor area used for storing or decanting chemicals must not be permeable. Old or corroded drums will cause more problems than those in good condition. Measures that can be taken to minimise contaminating the ground are:

- ☑ purchase chemicals in the appropriate sized containers to avoid the need for decanting
- ☑ where decanting is necessary, have safe procedures that avoid any spillage

Environmental management

- ☑ provide relevant information, instruction, training and supervision to employees
- ☑ dispose of all products in the correct way
- ☑ provide clear procedures and training for operatives to deal with accidental spillages
- ☑ make drip-trays available for plant in case it leaks environmentally damaging fluids
- ☑ have set procedures for the refuelling/replenishing of plant so that any spillage cannot permeate into the ground
- ☑ install bunding around all storage areas, even temporary fuel stores on construction sites
- ☑ maintain equipment or storage vessels in good condition
- ☑ get into the habit of only storing or using products that are needed, and only store such products in areas with impermeable floors without drain gullies
- ☑ maintain good housekeeping procedures and avoid the accumulation of litter or rubbish
- ☑ the burying of waste materials is strictly prohibited.

An emergency and incident response plan, appropriate to the size of the site and chemicals being used, should be in place in case of any spillages or pollution alerts (*see opposite*).

Be prepared – a site pollution response poster, spill kit and drip tray

Dealing with environmental incidents

While good planning can minimise the risk of an environmental incident, you should have a plan of what to do in the event of an emergency. Incident preparedness and response begins with considering what emergency scenarios and incidents may occur on a construction site and what in the surrounding environment may be impacted by them. Types of incident include fires, spills and leaks.

Environmental management

For each one of these types of incident several different parts of the environment may be affected. (For example, a fire will release products of combustion to atmosphere and firewater may enter the surface watercourse, trade effluent sewers and also percolate into the ground.)

Such a plan should include the following steps.

- ☑ **Stop.** How to stop pollution in the first place.
- ☑ **Contain.** How to keep the pollution in the area and stop it spreading any further.
- ☑ **Notify.** Who should be informed. This may include the regulatory agencies or emergency services as well as site management.
- ☑ **Clean-up.** How to clean-up the pollution and what to do with the wastes.
- ☑ **Train.** Make sure the site team know their roles and responsibilities and replenish the spill kit material.
- ☑ **Test.** Conduct a drill, ideally every six months, to reinforce the response needed.
- ☑ **Reporting.** You should report all environmental incidents to site management as soon as possible, or to the Hotline 0800 80 70 60 (24 hour freephone service).

Should you have an incident or be subject to an investigation on your site, environmental regulators have extensive powers to interview people, review documents including computer records and take copies. Regulators are being given powers to impose civil penalties for some minor offences, rather than reverting to the courts. Should you find yourself being investigated you should fully co-operate with the regulators and seek professional support.

Managing transport, materials, water and energy use

Transport

Vehicle movements and the transport of materials and people represent resources being used. It is important that vehicle movements are handled, programmed and managed in an efficient way as this will lead to greater resource efficiency and reduced costs.

Good logistics with prompt arrivals and departures with materials being off-loaded in the right locations will result in multiple positive benefits. The key benefits will be reduced nuisance, noise, dust, congestion, inconvenience and fuel use (and therefore less carbon emissions and associated cost savings). There is also less potential for double handling and wastage of materials.

Traffic management planning

A good traffic management plan will give consideration to the following factors:

- ☑ local traffic conditions, peak flows and congestion hot spots
- ☑ delivery and departure sequencing and times for all site vehicles
- ☑ off-site delivery and departure routes for all site vehicles
- ☑ lay down and delivery areas
- ☑ on-site traffic routes
- ☑ signage and directions on site
- ☑ hold areas for vehicles awaiting to off-load or depart the site
- ☑ communication between vehicles and logistics manager on and off site

Environmental management

- ☑ wheel-wash locations, dust suppression and mud sweeping
- ☑ receiving and responding to complaints
- ☑ consultation processes with emergency services, local residents, schools, public facilities and businesses.

Consolidation areas and sharing of transport

The use of off-site consolidation areas and the sharing of transport will reduce the number of vehicle movements to and from site. Vehicle sharing can be planned by site staff. Project suppliers may permit the use of their parking areas as car pooling points.

Parking and lay down areas

- ☑ On-site parking for staff, contractors and visitors needs to be clearly identified.
- ☑ Delivery vehicle and waste removal parking also needs identifying.
- ☑ Vehicle off-loading and lay downs areas should be clearly identified.

Plant and vehicles

Ensure that plant and vehicles are properly maintained to ensure efficiency of operations.

On-site traffic management (safe routes, dust and wheel wash)

- ☑ Use wheel washes to prevent mud and any contaminated materials getting onto local roads.
- ☑ Water bowsers can be used for dust suppression.
- ☑ Road sweepers can be used to remove any mud accumulations.

Materials

Construction uses huge amounts of natural resources and in the UK accounts for 25% of all raw materials used. Historically construction has been an inherently inefficient process, arising from the individual nature of on-site construction. This not only wastes a lot of money, it produces high levels of waste materials and causes excess material extraction to replace those materials that have been lost through inefficient use. It has been common practice to over-order by 5-10% to allow for site wastage from damage, spillage and theft, for example.

The aim of modern construction is to move to the top of the waste hierarchy (prevention) and away from where it has been traditionally placed, which is at the bottom (disposal).

Practical ways for improving resource efficiency and reducing waste include:

- ☑ design:
  - designing the project to incorporate recycled materials
  - designing the project to suit standard product sizes and to avoid site cutting
  - designing to allow pre-assembling components off site
  - designing to allow a cut/fill balance and by utilising surplus materials in site features (such as landscaping)
  - specifying non-hazardous and low impact materials
- ☑ procurement:
  - selecting suppliers with a good environmental track record
  - requiring sub-contractors to have a waste management policy
  - not over-ordering materials
  - reducing the amount of packaging
  - ordering materials at the size required, to avoid off-cuts

Environmental management

☑ **delivery and storage:**
- unloading carefully to avoid damage
- refusing to take delivery of damaged goods
- do not deliver to inappropriate areas of the site
- do not take delivery of incorrect deliveries (specification and quantity)
- do not exceed the shelf life of materials
- avoiding damage or contamination from incorrect storage
- keeping small/valuable items in a secure location
- storing waste in a designated area and in segregated waste streams
- storing material stockpiles to avoid silt run-off
- storing material stockpiles away from drains and watercourses
- storing oil, fuel and chemicals in bunded areas away from drains and watercourses

☑ **handling:**
- avoiding multiple handling of materials
- using the correct equipment to lift or move materials to avoid damage.

There are many benefits to using recycled aggregates and these help to reduce the demand for virgin materials, include lower embodied energy and less transport if produced on brownfield sites. On-site processing of demolition aggregates may be classed as a waste operation.

 For further information refer to GE 700 *Construction site safety,* Chapter E03 Waste management.

Water consumption

Whilst the dewatering from site excavations to avoid interference with construction activities does not need an abstraction licence, abstracting water for other uses (such as dust suppression, washing down and so on) will require a licence if using more than 20 m^3 per day in England and Wales.

Collecting rainwater for reuse

The Government's sustainability strategy for the construction sector identifies reducing water usage in the manufacturing and construction phase by 20% compared to a 2008 baseline (148 m^3 per £m of contractors' output). This target will both reduce pressure on water resources and also save money. There are a variety of techniques that can be used to increase water efficiency.

Environmental management

On larger construction projects, where high volumes of water are being used, the first action is to establish an approach to measuring and monitoring water usage so that it can be managed. This may involve the use of water meters at appropriate locations and the use of water balances to account for water usage. Water reduction targets can be set based on known volumes of water usage and progress monitored.

Significant savings can be made by using rainwater harvesting systems to collect rainwater from roofs and other flat surfaces. Early installation of suitable collection systems would need to be investigated at the design stage and payback times calculated for the expected volume of water use. Harvested water can be used for dust suppression, avoiding the need to draw water from the mains or abstraction.

During supervisors' site tours water use can be monitored and any obvious leaks and running hoses identified and dealt with. The use of triggers on hoses will prevent hoses from running whilst unattended.

Vehicle wheel wash is now available with water recycling and recirculation systems fitted. These will reduce the volume of water used and have the potential to save money. These systems work by providing a solids settling area combined with the use of flocculants to further precipitate solids out. The solids that are collected can be periodically removed.

Site accommodation can be fitted with waterless urinals, push taps and rainwater harvesting for toilet flushes.

Using a milk bottle to save flush water

Waterless urinals

Environmental management

Energy use and climate change

It is widely accepted that climate change is now seen as the defining challenge of this era. Whilst international negotiations are continuing to agree targets beyond the Kyoto Protocol, the UK has implemented the Climate Change Act, which sets out legally binding targets to reduce greenhouse gas emissions by 34% by 2020 and 80% by 2050 based on 1990 levels.

Many governments, including the UK, have taken action by introducing laws and regulations that specifically influence how we use energy (such as building regulations and the need for high levels of insulation in new build).

Some clients, building and construction companies now have to measure and report on their energy and fuel use and carbon emission through assessment schemes (such as BREEAM and the Code for Sustainable Homes). When requested through contract conditions or as good practice, observe good energy efficiency practices (such as switching equipment off when not in use) and monitoring and reporting where required (for example, fuel, electricity and gas consumption).

The Strategic Forum for Construction, together with industry leaders, has set out an action plan with a number of priorities that can be undertaken by companies to help reduce their energy and carbon emissions, such as:

- ☑ ensuring sites connect to the electricity supply as early as possible to prevent lots of equipment running on fuel
- ☑ installing more energy efficient site accommodation
- ☑ efficient use of construction plant (such as turning off when not required)
- ☑ good practice energy management on site:
  - plan transport/haulage to reduce part loads
  - avoid double handling or delivery to wrong areas
  - effective control of lighting/heating in welfare facilities
  - consider car sharing or crew buses
- ☑ making use of consolidation centres to reduce transport and handling of materials
- ☑ fuel efficient driving through driver training
- ☑ good practice energy management in company offices.

Plants and wildlife

Damaging, disturbing or removing protected species can result in prosecution under a range of environmental legislation. Wildlife is also often held in very high regard by the general public.

Construction activities have the potential to impact on the ecological environment. These impacts can be in the form of:

- ☑ disturbance of birds, bats, badgers and other protected species
- ☑ removal and fragmentation of habitats
- ☑ disturbance to aquatic wildlife and water quality
- ☑ disturbance to wildlife from noise and vibration
- ☑ damage to trees and hedgerows.

Protected species and plants

A number of species are specifically protected by legislation. If these are encountered on site then it is likely that a licensed ecologist will need to complete a survey and determine the best course of action.

Environmental management

Great crested newt

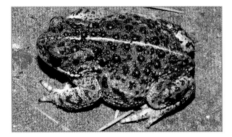
Natterjack toad

Bats

This could mean having to wait for the chicks of nesting birds to fledge and leave the nest.

Projects will have to work around any nesting birds as the nests, eggs and immediate habitat are protected. Examples of some protected species include:

- ☑ great crested newt
- ☑ water vole
- ☑ large golden ringed dragonfly
- ☑ natterjack toad
- ☑ bat
- ☑ otter
- ☑ smooth snake
- ☑ dormouse
- ☑ adder
- ☑ slow worm
- ☑ badger.

Some protected birds include:

- ☑ all wild birds
- ☑ common quail
- ☑ kingfisher
- ☑ peregrine.

Some protected plants include:

- ☑ fen orchid
- ☑ shore dock
- ☑ bluebell.

Please note the above lists are not exhaustive.

Environmental management

Invasive plants

Species of plants that do not naturally occur but have been introduced are known as **non-native** species. A number of these species have become invasive because native species are unable to compete against these bigger, faster growing and/or more aggressive introductions.

It is an offence to plant the following or otherwise cause them to grow in the wild:

- ☑ giant hogweed
- ☑ Japanese knotweed
- ☑ Himalayan balsam
- ☑ rhododendron
- ☑ floating pennywort
- ☑ parrot's feather
- ☑ Australian swamp stonecrop.

Giant hogweed

Japanese knotweed (image supplied by Bridget Plowright)

Himalayan balsam

Environmental management

The environmental statement for the project will highlight any concern for non-native species. If these are present then advice should be sought from specialists who will provide further details on how to best treat and dispose of these plants in each instance.

Before starting on site

The first step in dealing with any wildlife issues is to determine whether the client has identified any designated sites, protected species or invasive plants, together with any particular protection requirements.

This information will usually be contained within the contract documentation. Where a formal environmental impact assessment (EIA) has been carried out under planning requirements the environmental statement will highlight any protected species, habitats and areas, and any other environmental issues that may be relevant to the site. Contract or planning conditions may state, for example, that certain trees must remain undamaged.

Before any work commences all sensitive areas should be identified and fenced off, or access restricted, to prevent accidental damage. These issues should be addressed in the project environmental management plan (EMP) and communicated to site personnel through induction, briefings, toolbox talks and work method statements.

Careful planning should take place to ensure that work on areas of vegetation destined for removal are done outside of any breeding or nesting season to avoid damage or disturbance to protected species.

A regular inspection regime should be put in place to ensure that all protection to sensitive areas is being maintained in a satisfactory condition without damage or encroachment by the construction works.

Archaeology and heritage

Archaeological remains and the built environment provide a valuable record of a nation's history and identity and are an irreplaceable part of its national heritage. For these reasons archaeology and built heritage form a key element of planning policy and must be considered early in any construction project, particularly those which require a formal EIA.

Suitable controls to manage archaeology and heritage should be included in the project as early as possible to:

- ☑ ensure buildings are designed to avoid the disturbance of remains or historic features
- ☑ avoid disturbance during the construction process itself
- ☑ enable designers to incorporate historic features of the site in the final development
- ☑ comply with legal requirements relating to scheduled monuments, listed buildings and protection of the historic environment.

Protected historic remains

There are a number of different types of historic remains that are given legal protection, including:

- ☑ scheduled monuments (requiring a scheduled monument consent to work on or in the vicinity)
- ☑ human remains and burial grounds (unexpected remains must be reported to the coroner)
- ☑ treasure (coins at least 300 years old, objects containing at least 10% gold and anything found in the vicinity of known treasure)
- ☑ listed buildings (historic buildings are given greater protection)
- ☑ parks and gardens (historic parks and gardens can be registered).

Environmental management

Before starting on site

The environmental statement and planning conditions for the project provided by the client should identify any obligations for the management of archaeology and heritage. These requirements should be incorporated into the EMP and method statements. In some cases it will be a condition of the planning consent to prepare an archaeological management plan, which may include the employment of a qualified archaeologist as a watching brief during relevant construction works.

Areas of known archaeological or historic interest should be protected with suitable fencing to prevent damage or encroachment.

It will also be important to agree the appropriate method of working adjacent to sensitive areas, as vibration from operations (such as excavation or tunnelling) may cause damage. There may be an obligation to provide vibration monitoring of the works to ensure that vibration levels are not exceeded.

De-watering schemes may also have an impact on archaeological features as this could cause differential settlement or damage to materials that have previously been protected by being waterlogged. Appropriate methods of de-watering should be agreed in advance of the works taking place.

Unexpected finds

If any unexpected or suspected finds are encountered, work should stop in the affected area and be fenced off to protect it.

Where there is no archaeological watching brief employed by the project, advice should be sought by contacting the Local Authority archaeological officer on how to proceed. If human remains are found it will be necessary to report this to the coroner and for authorisation to continue approval will need to be given by the Home Office.

Land use and contamination

Building and construction work often involves the redevelopment of land previously used for commercial or industrial activities. These are often called brownfield sites. The surface of the ground itself and the ground beneath the surface may be contaminated by materials that have been worked, stored, spilt, buried, dumped or abandoned on the land in previous years.

This will also include the residue, waste or by-products from some industrial processes and the ashes from fires. Both solid and liquid waste may have permeated into the ground to a considerable depth.

Sites with previous industrial occupation should be assumed to be polluted, and tests undertaken to ascertain the types of pollutant and their concentration.

Examples of previous industrial activity, together with the likely chemical contamination are:

- [✓] **oil refineries and gas works** (fuel oil, lubricants, bitumen, alcohols, organic acids, PCBs, cyanides, sulphur, vanadium)
- [✓] **lead works** (lead, arsenic, cadmium, sulphides, sulphates, chlorides, sulphuric acid, sodium hydroxide)
- [✓] **pesticide manufacturing** (dichloromethane, fluorobenzene, acetone, methanol, benzene, arsenic, copper sulphate, thallium)
- [✓] **textile and dye works** (aluminium, cadmium, mercury, bromides, fluorides, ammonium salts, trichloroethene, polyvinyl chloride).

The assessment of contaminated land is very complex and should be carried out by competent persons. Everyone involved in work on such land must make an assessment of potential risks to human health and the environment, and implement any protective measures that need to be taken.

Environmental management

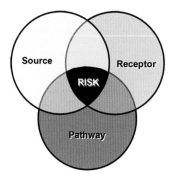

Dealing with contaminated land is a three-stage process.

- ☑ **Stage 1.** Identify the risks by reviewing the source contaminants, pathways and receptors.
- ☑ **Stage 2.** Decide upon the most appropriate remediation strategy to deal with the contamination.
- ☑ **Stage 3.** Implement the remediation works on site.

A duty on clients under the Construction (Design and Management) Regulations is to provide all contractors with relevant health and safety information on the project, which in appropriate circumstances will include details of the previous use of the land (going back historically as far as is practical) and any buildings or structures (existing or now demolished) that were built upon it.

 If working on a site and you come across unusually coloured soils, liquids, sludge or encounter any unusual smells or odours – cease work immediately, evacuate the area and seek professional advice.

Declaration of a clean site

The management responsible for workings on the site will formally decide when a site is free from contamination, and declare this so any necessary fencing and decontamination facilities previously provided can be removed.

Occupational health considerations

The health of workers on contaminated sites can be affected through one or more of the following ways:

- ☑ asphyxiation
- ☑ gasing
- ☑ ingestion
- ☑ inhalation
- ☑ skin absorption
- ☑ skin penetration.

Suitable personal protective clothing and approved respiratory protective equipment must be worn at all times when work is carried out on contaminated sites.

Personal hygiene

The level of risk to health by any contaminants will determine the scale of need for hygiene facilities, but certain considerations (such as the provision of a drench shower), should always be borne in mind when working on a contaminated site.

- ☑ A dirty area is required for workers to discard dirty or contaminated clothing. Such clothing should be bagged and identified within this area before being dispatched to specialist cleaners.

Environmental management

☑ A clean area is required for workers to put on clean and non-contaminated clothing. Access to and exit from this area must only be to the clean part of the site. It is essential that the entry/exit point of the clean area is in the clean part of the site.

☑ Toilets, showers and washing facilities should be positioned between the dirty and clean areas, so that workers may wash or shower in order to remove any contaminant from their bodies.

 For further information refer to Chapter E24 Waste management, and GE 700 *Construction site safety*, Chapter E03 Waste management.

Environmental management

24

Waste management

What your employer should do for you	346
What you should do as a supervisor	347
Introduction	348
Waste regulation authorities	349
Understanding and describing waste	350
Waste priorities – reduce, reuse and recycle	351
Waste duty of care	352
Waste management permits, licenses and exemptions	353
Managing waste	354
Disposal of waste off site	360

Waste management

What your employer should do for you

1. Employ a competent member of staff who is familiar with waste management and waste legal requirements that will take overall responsibility for waste on the project.
2. Ensure that all relevant staff receive a site waste management plan induction/briefing so that they are aware of the correct on-site procedures for handling, storing and recording waste.
3. Prepare a site waste management plan (SWMP), where appropriate, identifying the project wastes and how they are minimised and managed and sign a joint declaration with the client.
4. Engage the design team to take steps to minimise waste during the design process if relevant.
5. Ensure that all proposed disposal facilities have a valid waste management permit or licence.
6. Ensure that all proposed waste carriers have a valid waste carrier licence.
7. Ensure that sub-contractors and suppliers are aware of the project waste requirements and that they contribute to the SWMP.
8. Ensure the SWMP is regularly reviewed and updated and that current copies are available.
9. Register the site as a hazardous waste producer if plans are to produce more than 500 kg of hazardous waste.
10. Liaise with the waste regulators as appropriate and apply for relevant waste management permits and exemptions.
11. Maintain records of the SWMP and associated waste transfer documentation – two years for non-hazardous waste and three years for hazardous waste.

Waste management

What you should do as a supervisor

Checklist	Yes	No	N/A
1. Liaise with all sub-contractors and suppliers under their control to ensure the site waste requirements are being fulfilled.			
2. Regularly monitor waste containers and skips to ensure they contain the correct materials.			
3. Ensure that all work areas under their control are safe and free of debris.			
4. Ensure that all skips and containers are properly identified for the waste they are allocated.			
5. Notify any waste infringements to the relevant personnel.			
6. Arrange for skips and containers to be regularly emptied.			
7. Ensure that waste transfer documentation is being completed correctly before the waste leaves the site.			
8. Ensure that waste transfer documentation is returned to the site office for inclusion in the site waste management plan/site records.			
9. Engage in regular site tours and inspections with site staff and contractors to review current waste practices and performance.			
10. Carry out regular toolbox talks and briefing on waste as and when appropriate.			

Waste management

Introduction

The construction industry is the single biggest consumer of resources in the UK and consumes around 420 million tonnes of materials per year. The industry also generates a huge amount of waste – around 120 million tonnes a year; about 20 million tonnes still goes to landfill, including 10 million tonnes of new, unused building products. The UK Government sees the reduction of construction waste as a strategic issue that also contributes to its climate change agenda. The UK Government's strategic targets for construction waste is to halve all construction waste to landfill by 2012 and to have zero waste to landfill (based on cost) by 2020. This will create significant challenges and opportunities for the construction industry.

The poor management of materials and resources on a construction project can lead to excessive amounts of waste, which is costly, is bad for the environment and can be unsafe. Waste can be generated through a number of reasons including poor design, incorrect or over-ordering, poor workmanship, incorrect storage and management leading to damage and theft of valuable items. The typical cost of waste on a project is around 5% of the build cost.

The following are the key points covered in this chapter:

- ☑ the improper disposal of waste is illegal and can lead to prosecution and even imprisonment
- ☑ producers of waste must correctly identify their waste as inert, non-hazardous or hazardous
- ☑ in England and Wales, producers of 500 kg or more hazardous waste (such as oils or asbestos), must register their premises with the Environment Agency (in England) or Natural Resources Wales
- ☑ producers of waste have a legal duty of care to ensure that it is passed on to an authorised person with the correct technical competence
- ☑ all contractors who carry or collect waste should have a waste carrier's licence
- ☑ all waste disposal facilities should have a waste management permit (England and Wales) or licence (Scotland) unless they have a registered exemption from the Environment Agency, Natural Resources Wales or Scottish Environment Protection Agency
- ☑ all waste transfers must be supported by the correct document, called a controlled waste transfer note (the transfer of hazardous waste requires a consignment note)

Waste management

☑ from April 2008, projects in England with an estimated value of £300,000 and above are required to prepare a site waste management plan, identifying the types and quantities of waste and actions to reduce these wastes. A more detailed plan is required for projects above £500,000.

Waste regulation authorities

In England the Environment Agency (EA) is the main authority for enforcing waste legislation and issuing permits and exemptions. In Wales it is Natural Resources Wales and in Scotland it is the Scottish Environmental Protection Agency.

Local Council Authorities also have some waste enforcement powers, such as the stop and search and seizure of vehicles suspected of waste crime (such as fly tipping) and for lesser offences (such as littering). Local Authorities are also responsible for issuing waste permits involving air pollution control (such as mobile crushing equipment).

Waste responsibilities

There will be a number of personnel involved in a construction project that have key responsibilities for waste management. For a typical construction project requiring a site waste management plan, the table below provides an example of the waste responsibilities for the project team.

Project member	Waste management responsibility
Client	■ To ensure the SWMP is started before the works commence. ■ Setting waste standards and targets in the project specification. ■ To regularly review and sign off changes to the SWMP.
Project manager	■ Ensuring the client is aware of their legal responsibilities with respect to the SWMP and asking them to sign/approve each version of the SWMP. ■ Ensuring that a suitably competent person is employed who is familiar with waste management activity and legislation. ■ Ensuring that the SWMP is started as early as possible in order that waste can be minimised through design and procurement. ■ Setting an expectation of high waste management standards. ■ Providing necessary leadership to ensure its implementation, and ensuring co-operation from the rest of the team. ■ Ensuring that the project team and contractors are engaged in the development and implementation of the SWMP by ensuring its implications are embedded in sub-contracts and method statements, discussed in meetings with sub-contractors and site team, is a topic of focus on site inspections/walk rounds, and is appropriately policed. ■ Taking responsibility for having all required waste-related documentation. ■ Producing and promoting the site waste rules in co-operation with the person in charge of the SWMP.

Waste management

Project member	Waste management responsibility
Design manager	▪ Contributing to measures to design out waste. Helping to co-ordinate design with construction, ensuring that waste minimisation plans are sustained and implemented.
Commercial manager	▪ Contributing to waste minimisation and waste management actions and ensuring quantities reflect the waste minimisation actions.
Procurement manager	▪ Ensuring that relevant actions for waste minimisation and recycling are included in relevant tender documents and contracts. Ensuring that all contractors are invited to submit their own ideas for waste minimisation.
Construction supervisor	▪ Contribute to waste minimisation and waste management actions. Ensure co-ordination of construction process so that all waste minimisation and management actions are implemented.
Engineering staff	▪ Support construction supervisor with above actions, particularly with work packages overseen by the engineer.
Representatives of any sub-contractors who provide own waste skips/bins	▪ Be engaged fully in site waste management.

Understanding and describing waste

Waste generally means a substance or object which the holder discards, intends to discard or is required to discard, although many court cases in the UK and Europe highlight that whether or not the substance is a waste may be subject to review in each particular case. This is because a material can be classed as waste even if it has a use or someone is prepared to pay for it.

Waste can be divided into three types.

- ☑ **Inert (inactive).** Waste that will not decompose to produce greenhouse gases (such as bricks, concrete, tiles, ceramics and glass).
- ☑ **Non-hazardous (active).** Waste that will decompose but does not contain dangerous substances (such as timber, food and paper).
- ☑ **Hazardous (active).** Waste that contains dangerous substances and is dangerous to human health or the environment or as stated in the list of wastes.

It is your responsibility as the producer of waste to:

- ☑ decide what is and is not waste
- ☑ record and retain details (in the SWMP if applicable).

Waste management

Cardboard and shrink wrap packaging waste. Consider using reusable or returnable packaging to avoid cost of waste removal

Waste transfer documentation must contain information that properly describes the waste so that waste disposal sites know what type of waste they are accepting. Across the UK there is a standard system of waste codes called the list of wastes (LOW) in England and Wales, and the European waste code (EWC) in Scotland, which must be put on the controlled waste transfer note or the hazardous waste consignment note.

Under the LOW/EWC, each waste has a six-digit code. The first two digits refer to the activity that produced the waste, the second two digits refer to the sub-chapters for the different waste types and the final pair of digits correspond to the individual waste stream.

 17 01 02 is the LOW/EWC code for bricks, as follows:

- ☑ 17 refers to construction and demolition waste
- ☑ 01 refers to concrete, bricks, tiles and so on
- ☑ 02 refers to bricks.

Any waste code listed within the LOW/EWC and marked with an asterisk (*) is considered as a hazardous waste and should be managed accordingly.

 It is not acceptable just to enter 'muck away' or 'general waste' on the waste transfer note as it is a legal requirement to include a description and the six-digit waste code.

Waste priorities – reduce, reuse and recycle

When thinking about how to deal with waste there are a number of options, with each option having different levels of environmental benefit. The waste hierarchy, shown on the diagram below, identifies the various options from prevention (most preferred) to disposal (least preferred). The well-known phrase of 'reduce (prevent or minimise), reuse or recycle' comes from various levels within the waste hierarchy.

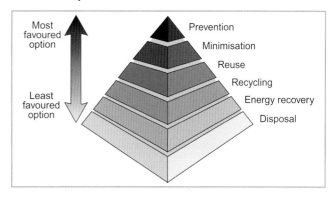

Waste management

The following table includes examples of waste minimisation principles at different stages of a construction project.

Design	Procurement	On-site processes
▪ Use a smaller number of component types (for example, standardised doors and window). ▪ Avoid unnecessary material use. ▪ Set dimensions to minimise off-cuts. ▪ Design to enable maximum application of procurement and on-site waste minimisation principles. ▪ Integrate demolition and build into one project and maximise reuse of demolition material.	▪ Manufacture off site with reusable packaging of all delivered units. ▪ Eliminate packaging. ▪ Avoid unnecessary layers of packaging. ▪ Reduce thickness of packaging. ▪ Take-back of packaging, excess product and damaged product. ▪ Take-back of pallets. ▪ Include contractual clauses to penalise waste production/incentivise waste reduction or require waste-reducing activities. ▪ Include contractual clauses to require recycling/reuse.	▪ Repair pallets or use plastic, reusable pallets so no new pallets arrive on site. ▪ Arrange bulk deliveries (but be aware of handling implications). ▪ Arrange just-in-time deliveries. ▪ Provide a good, recognised storage area for reusable products on site. ▪ Maximise segregation.

 The Waste (England and Wales) Regulations require that you declare on the waste transfer documentation that you have considered the waste hierarchy in the management of the waste.

Waste duty of care

Under duty of care all waste must be properly stored (to prevent escape), transferred and ultimately disposed of in a responsible manner. All transfers of waste must be made to properly authorised persons holding the relevant permit or licence.

In England and Wales there is a two-tier system of waste carriers; but those involved in transporting construction and demolition waste fall into the upper tier and must register a waste carrier, together with paying the relevant fee, **even if you carry your own construction waste.** There are also requirements to register where you are a waste broker (those who arrange the movement/disposal on behalf of others) and waste dealers (those who use an agent to buy and sell waste). It is possible for a single registration to cover all these activities. The relevant form can be obtained from the Environment Agency's and Natural Resources Wales' websites.

Similar rules apply in Scotland with the relevant form being obtained from the local waste regulation office.

Waste management

The duty of care also means that waste must be transferred to a waste transfer station or disposal site holding the right permit or licence, together with appropriate documentation in the form of a controlled waste transfer note (for non-hazardous waste) or a consignment note (for hazardous waste).

A building contractor or anyone else who carries waste (such as a plumber, carpenter, bricklayer or decorator) must be registered as a carrier even if they carry only their own waste. They must have the necessary documentation to consign the waste to the management of a licensed site who will, in turn, acknowledge receipt and give details in writing of where and when the waste is to be disposed of.

Carrying surplus materials back to your workshop or stores for sorting and reuse does not constitute the carriage of waste.

Waste management permits, licenses and exemptions

The Environmental Permitting (England and Wales) Regulations specify which waste activities require an environmental permit and allow some waste activities to be exempt from requiring a permit (in Schedule 3) as shown below. An environmental permit is required for five types of waste operation.

- ☑ **A waste installation.** A waste facility carrying out any activity listed in Schedule 1 of the regulations, including landfill sites and waste transfer stations.
- ☑ **A waste operation.** The depositing, treatment or recycling of waste that is not exempt under the Environmental Permitting Regulations, including the treatment of contaminated land with treatment plant.
- ☑ **Mobile plant.** Mobile equipment carrying out an activity listed in Schedule 1, or a waste operation (for example, mobile soil treatment plant and crushers).
- ☑ **A mining waste operation.** The management of mining extraction waste, which may include the mining waste facility.
- ☑ **Radioactive substances.** The keeping and management of radioactive material, including radioactive apparatus, or the storage and disposal of radioactive waste.

It is an offence to operate any of the above waste management activities without a permit or registered exemption. Exemptions refer to specific low-risk waste handling operations that do not require permits and have strict compliance criteria.

Some wastes can be stored under an exemption, including:

- ☑ recycling and reuse of materials
- ☑ using the waste as a construction material.

As part of introductory discussions with your waste regulators *(see 'Managing waste' on the next page)* you should discuss whether exemptions apply to your project. Applications are required for most exemptions under one of four categories.

- ☑ **Use.** Operations with direct use of material (for example U1 – Use of waste in construction).
- ☑ **Treatment.** Where the waste is changed in some way (for example T5 – Screening or blending of waste to produce an aggregate or soil).
- ☑ **Disposal.** Where waste is permanently destroyed or deposited (for example D1 – Deposit of waste from dredging of inland waterways).
- ☑ **Storage.** Pending recovery elsewhere (for example S2 – storage of waste in a secure place).

Exemptions are relatively simple and quick to obtain.

Waste management

Where your operations do not fall into the strictly specified conditions for exemptions it is likely you will need either a standard or a bespoke permit. Such permits can take several months to obtain and carry significant competence and cost implications.

In Scotland, similar rules apply except that environmental permits are called environmental licenses and exemptions are either simple or complex, depending on the activity concerned.

Under certain circumstances it is possible to reuse recycled waste on site without the need for exemptions from environmental permitting. Special provisions are set out in WRAP Quality Protocol and CL:AIRE Definition of Waste: Development Industry Code of Practice (DiCOP). Under these provisions, wastes are treated in a licensed or permitted facility (either on or off site) and then managed as non-wastes and can be reused on site. The materials must be controlled under a materials management plan (MMP) under the direction of a qualified person.

If you propose to follow this route you must make yourself familiar with the exact requirements about moving waste and receiving non-waste, as well as completing checks on the source of materials brought onto your site. It is important to liaise closely with the waste regulator to see if this is permissible on your site.

If you believe you require an exemption or a permit then you should speak to your local waste regulator at the earliest opportunity.

The detailed regulations and guidance are freely available on the regulators' websites.

Managing waste

Before starting on site

It is best practice to carry out the following steps at the start of the planning phase for any construction work (see also CDM definitions). It may be necessary to obtain exemptions or permits from the regulator that can take some time to obtain. Early discussions can prevent unexpected delay.

- [✓] Obtain the contact details of the waste regulator and Local Authority in whose area you will be working.
- [✓] Contact the waste regulator and discuss as appropriate any plans for the treatment and recovery of any on-site materials.
- [✓] Discuss how and where you intend to move the waste and who will be moving it.
- [✓] Consider options for handling the materials as non-waste.
- [✓] Investigate and share options for on or off site treatment plans.
- [✓] Check that any waste carriers are licensed and listed on the public records.
- [✓] Check any waste transfer, treatment, recovery or disposal sites are correctly permitted and listed on public records.
- [✓] Review the need for any waste exemptions or permits that might be needed.
- [✓] Where required (see below) the SWMP for the project should be completed as far as practicable before the commencement of work.

Waste management

Site waste management plans

Since April 2008, projects in England with an estimated value of £300,000 (excluding VAT) and above will need to prepare a site waste management plan (SWMP), to identify the types and quantities of waste that will be produced and the actions that will be taken to reduce these. A more detailed plan is required for projects in excess of £500,000. The purpose of the plan is to reduce the amount of waste on site, encourage reuse and recycling, and prevent fly-tipping. It is good practice to have a SWMP for all but the very smallest of projects. Typically a site waste management plan will:

- ☑ enable consideration of waste generation and options before a project starts
- ☑ document how waste has been designed out of a project
- ☑ identify who is responsible for resource and waste management on the project
- ☑ show what types of waste will be produced
- ☑ detail how the waste will be managed (reduced, recycled, reused or disposed of)
- ☑ identify which contractors will be used to handle the waste
- ☑ set out how the waste quantities will be measured.

In addition to being a legal requirement, implementing a SWMP can bring a number of business benefits:

- ☑ wastes can be identified early and minimised through design and procurement practices before construction starts
- ☑ queries from the waste regulator can be answered simply and easily

- ☑ it helps to avoid prosecution by ensuring that all wastes being disposed of end up in the right place
- ☑ it shows how waste is managed and helps to reduce costs; materials and waste are managed responsibly and are therefore of less risk to the environment
- ☑ it helps to provide valuable information for future projects on the costs and quantities of waste produced. This information can be used to set targets for reduction.

Responsibilities under site waste management plans

What the client must do

- Produce the initial SWMP before construction work begins by using the knowledge of the design team as necessary.
- Appoint the principal contractor.
- Pass the SWMP to the principal contractor.
- Continue to have a role in ensuring effective implementation.
- Conform to the declaration that all waste has been dealt with in line with legislation.

Waste management

What the principal contractor must do

- Obtain relevant information from contractors.
- Update the SWMP as often as necessary as work progresses and not less than every six months.
- Keep the SWMP on site.
- Ensure contractors and the client know where the SWMP is kept and have access to it.
- At the end of the project (not later than three months), review the plan, record reasons for any differences and confirm that the plan was monitored.
- Hand the SWMP back to the client.
- Keep a copy of the SWMP for two years.
- There are additional duties for projects over £500,000.

What contractors must do

Check the contract for requirements on:

- purchasing strategies or methods of work aimed at reducing waste
- the on site reuse or recycling of site gained (generated) materials
- the disposal of waste
- what information has to be reported to the principal contractor or client and when.

The level of detail that the SWMP should contain depends on the value of the contract excluding VAT.

For projects estimated to be between £300,000 and £500,000 the SWMP should contain the:

- ☑ types of waste removed from site
- ☑ identity of the person who removed the waste from site
- ☑ site(s) that the waste was taken to.

For projects estimated at over £500,000 the SWMP should contain the three points above, plus the:

- ☑ waste carrier registration number
- ☑ environmental permit or exemption held by the site(s) where the material was taken
- ☑ reference to, or the written description of, the waste.

Because the SWMP is produced prior to the start of the project (that is, pre-start site conditions must be taken into account), all parties have the opportunity to develop ways of reducing waste, including the reuse or recycling of site gained materials as part of the project.

By identifying at an early stage the waste that cannot be reused, the possibility of alternative uses may be considered.

The SWMP must be the master document recording all the waste management for the site.

Managing hazardous (special) waste

Hazardous waste contains substances in sufficient quantities that make it dangerous to human health or the environment. In Scotland hazardous waste is called special waste.

Waste management

In England and Wales, all construction premises that produce 500 kg or more of hazardous waste, in any 12-month period, must be registered with the Environment Agency or Natural Resources Wales before the waste is removed from site. This can be done either on a paper form or electronically. You should contact the local agency's hazardous waste registration team to register your premises. Registration must be renewed annually. In Scotland there is no requirement to register.

The following controls should be adopted to comply with the hazardous or special waste regulations.

- ☑ Different types of hazardous wastes should be segregated as you will clearly need to identify the quantities and types on the hazardous waste consignment note (see below).

- ☑ Mixing of different types of hazardous waste should be avoided as this may inadvertently create an explosive or fire risk, particularly in warm weather.

- ☑ Mixing of hazardous waste with non-hazardous waste to dilute the material below the threshold concentration is banned.

- ☑ Packaging or containers contaminated with hazardous substances should be treated as hazardous waste unless it can be shown that the concentration (including the packaging) is below the threshold limits.

Where individual products are combined to form a substance (for example, adhesives and resins), then each component should be considered for its hazardous properties and disposed of accordingly. Resins are often inert when set so leaving materials to dry before disposal may make them non-hazardous.

Treatment of waste

The Environmental Permitting (England and Wales) Regulations and the Landfill (Scotland) Regulations state that non-hazardous waste must now be treated before being sent to landfill.

Treatment in this respect means that the waste must satisfy all three criteria of a three-point test. The treatment must:

- ☑ be a physical, thermal, chemical or biological process (including sorting)

- ☑ change the characteristics of the waste

- ☑ be carried out in order to:
  - reduce the volume of the waste, or
  - reduce the hazardous nature of the waste, or
  - facilitate handling of the waste, or
  - enhance recovery of the waste.

On construction sites, in practical terms, this can be achieved by setting up appropriate segregation skips and separating out (in other words sorting) any wastes that can be reused or recycled, which will change the characteristics of the original waste stream. This in turn will aid in reducing the volume of the waste, facilitating the handling of the waste and enhancing recovery of the waste destined for landfill.

Hazardous wastes are required to be stored and disposed of separately from non-hazardous wastes.

E24

Waste management

Compaction of waste is not considered as pre-treatment on its own. If the waste has been pre-sorted or segregated, and then compacted, this is acceptable. On construction sites, in practical terms, this can be achieved by setting up appropriate segregation skips and separating out (sorting) any wastes that can be reused or recycled, which will change the characteristics of the original waste stream. This, in turn, will assist in reducing the volume of waste, facilitating the handling of waste and enhancing recovery of waste destined for landfill.

A colour coded scheme can make the accurate segregation of waste easier.

Use of skips or containers

When deciding on the type and number of containers you require consider the type of waste that will be produced (this should be done at SWMP stage).

- ☑ Would a compactor skip be better than an open one?
- ☑ Would a compactor skip cut down on the number of skips you require?
- ☑ Would a tailgate skip be easier to use than a fixed one?
- ☑ Is the skip the correct size and will it hold all your wastes (prevent wind-blown litter, scavenging or leaking)?
- ☑ How many skips will you need and how often will they need to be emptied?
- ☑ How many different types of waste will you produce? For example:
  - brick rubble, concrete and cement
  - gypsum or gypsum-related products
  - empty paint tins, adhesive tins or drums
  - wood, cardboard, paper and carpeting
  - scrap metal (such as pipes or wires)
  - various chemicals, oils and greases
  - PVC window mouldings, gutters and downpipes.

Dispose of waste appropriately

Remember

- ☑ Place skips where contractors or your carrier's lorries can reach them easily.
- ☑ Keep all access to skips clear.
- ☑ Do not overload any skip; if you do, your carrier has every right to tell you to unload it, or even refuse to take it away.
- ☑ Do not load the skip above the height of the sides.
- ☑ Make sure that all waste is stored in the skip and not spilled around it.
- ☑ Do not light fires in skips.

Waste management

- [✓] Never allow anyone to climb into or ride in a skip (it is a dangerous practice and could result in waste being inadvertently tipped on top of them).

- [✓] Consider whether any, or all, of the skips need to be covered to prevent contaminated water run-off.

- [✓] Avoid storing more than 50 m^3 capacity at any one time, above this and you may have to register for an exemption or permit.

Managing waste electrical and electronic equipment (WEEE)

Waste electrical and electronic equipment (WEEE) comes under the WEEE Regulations. If you are considering the disposal of waste electrical equipment, the following actions must be taken.

- [✓] It must be segregated from other types of waste for disposal.

- [✓] If the waste electrical equipment was purchased before 13 August 2005, and is being replaced with new equivalent equipment, then ask the producer for details of their producer compliance scheme and collection arrangements.

- [✓] If the waste equipment is not being replaced with new equivalent equipment, or the producer compliance scheme cannot be traced, then you must pay to transfer the waste equipment to an approved authorised treatment facility that can accept waste electrical equipment (that is, authorised transfer station).

- [✓] If it was purchased after 13 August 2005, then contact the supplier for details of the producer compliance scheme and collection arrangements (if these have not been provided).

- [✓] Any waste transferred to an authorised collector or waste carrier must meet all of the normal requirements for duty of care, in other words waste carrier's licence, transfer notes and licensed treatment facilities (that is, waste transfer station) approved by the waste regulation authority.

Managing waste batteries

The construction industry is a large user of batteries in many types of vehicles, plant and equipment. In 2010 the recycling rate for portable batteries was low at around 3%.

The regulations deal with three types of battery.

Automotive batteries – used for starting, or the ignition of, a vehicle engine, or for powering the lights of a vehicle.

Industrial batteries – used for industrial or professional purposes (such as the battery used as a source of power and propulsion to drive the motor in an electric forklift).

Portable batteries – that are sealed, can be hand carried and are neither an automotive battery or accumulator nor an industrial battery. Examples of a portable battery include the AA or AAA batteries used to power a small hand-torch or the battery used to power a laptop or mobile telephone.

Take-back of waste batteries

Distributors of **portable batteries** (for example, retail stores), have a duty to take back waste portable batteries through facilities such as in-store waste-battery bins. This requirement does not apply to distributors who supply fewer than 32 kg of batteries per year.

A producer of **industrial batteries** is obliged to provide for the take-back of waste industrial batteries free of charge from the end user where:

- [✓] the producer has supplied new industrial batteries to that end user

- [✓] for any reason, the end user is not able to return waste industrial batteries to the supplier who supplied the batteries, providing the waste batteries are the same chemistry as the batteries the producer places on the market

Waste management

- the end user is not purchasing new batteries and a battery with the same chemistry as the one being returned has not been placed on the market for a number of years, then the end user's entitlement is to be able to contact any producer to request take-back.

The regulations also require that **automotive battery** producers collect, on request, waste automotive batteries free of charge, from businesses such as garages, scrap yards, end-of-life vehicle authorised treatment facilities or civic amenity sites (such as Local Authority waste recycling centres), during any calendar year in which the producer places new automotive batteries on the market. Under the regulations producers do not have a duty to collect waste automotive batteries from individual end users.

The burning of waste

During construction operations, especially demolition, the burning of waste is sometimes considered. There are very few situations where burning of waste is a permitted and legal option. In most instances burning of waste on site is banned through company rules.

 For further information on bonfires on site refer to Chapter C15 Fire prevention and control.

Difficult wastes

Although not legally defined, there are some types of waste that need to be handled in a different way. This is because there are specific legal controls in place for that waste or that handling of it at the landfill site might be problematic (for example, dusty materials).

In particular, gypsum and plasterboard wastes (or other wastes with high sulphate content), from 1 April 2009, can no longer be sent to landfill mixed with biodegradable waste. It must be separated from all other waste and disposed of to landfill separately or recycled. Many suppliers of plasterboard now offer recycling services; this should be considered ahead of landfill.

Other difficult wastes include:

- **invasive plants** – waste materials, both soil and plant matter contaminated with Japanese knotweed or Giant hogweed, can only be disposed of at sites that are specifically licensed to receive them

- **contaminated soil** – is that which is a mixture of soils, stones, rubble and polluting substances and could be a range of things left over from former use of the site.

Disposal of waste off site

Step 1 – Classifying the waste

It is a legal requirement for waste disposal sites to have a clear description of any waste they are accepting. This should be annotated on the duty of care waste transfer note or the hazardous waste consignment note, as highlighted above, the description of your waste can be identified using a system called list of wastes (LOW) in England and Wales and the European waste catalogue (EWC) in Scotland.

Under the LOW/EWC, each waste has a six-digit code – the first two digits refer to the activity that produced the waste, the second two digits refer to the sub-chapters for the different waste types and the final pair of digits corresponds to the individual waste stream.

 17 01 02 is the LOW/EWC code for bricks, as follows:

- 17 refers to construction and demolition waste
- 01 refers to concrete, bricks, tiles and so on
- 02 refers to bricks.

Waste management

When you carry waste for disposal, or contract a carrier to dispose of it for you, you must give an honest description of the waste, including the six-digit waste code.

Step 2 – Handing over the waste

Under duty of care you must pass on your waste to an appropriately authorised person. That is someone holding a valid waste carrier licence if the waste is collected or a waste transfer station or landfill site holding a valid environmental permit or licence if you are taking the waste yourself to the disposal site.

Waste transfer documentation

Whenever you pass the waste on to someone else (even if you carry your own waste) it must be supported by a waste transfer note (for inert and non-hazardous wastes) or a consignment note (for hazardous waste).

A controlled waste transfer note must:

- [✓] identify the producer of the waste
- [✓] give a description of the waste, including the six-digit EWC code
- [✓] state the quantity
- [✓] state how the waste is contained, whether loose or in a container and, if in a container, the kind of container
- [✓] give the place of transfer from your ownership into the waste site management
- [✓] state the date and time of transfer
- [✓] contain your signature and the signature of the authorised person receiving your waste.

In England and Wales whenever you pass waste on to someone else, you will have to declare on the waste transfer note, or consignment note for hazardous waste, that you have applied all reasonable measures to apply the waste management hierarchy. It is also a requirement to include the appropriate 2007 Standard Industry Classification (SIC) code on all waste transfer notes. The existing hazardous waste consignment note will continue to use the 2003 SIC code.

Step 3 – Disposal of waste

At the point of disposal (such as the waste transfer station or landfill site) the receiving facility will sign the controlled waste transfer note or consignment note and will issue a receiving ticket. For traceability purposes regular checks of this documentation should be made to verify the destination that was originally identified when the waste left site. This information will also be useful for completing the SWMP and other monitoring requirements, particularly where the project is being assessed under BREEAM.

Removal of contaminated waste

Where contaminated waste and other materials are to be removed from a site, protective sheeting for skips and lorries will be necessary.

All skips or vehicles must be completely sheeted within the dirty area of a site. Care must be exercised by those carrying out the sheeting operations to ensure they do not come into contact in any way with contaminated materials.

Vehicle drivers should not sheet their own vehicles, except to finally tighten sheet ropes, which should only be done in the clean area of the site.

Waste management

Facilities must be available to thoroughly wash all vehicles leaving a contaminated site. Detailed records must be kept of the disposal of hazardous or contaminated waste.

Other means of disposal

It is illegal to dispose of any liquid waste, including paint and solvent residues, by pouring it into drains or allowing it to soak into the ground. The environmental impact will likely be significant. Such wastes are likely to be classed as hazardous and must be disposed of in accordance with current waste disposal legislation.

Solid waste (such as brick rubble, off-cuts of roofing felt and other scrap materials) must not be buried. The different types of waste should be segregated into separate skips. From the perspectives of site tidiness and health and safety, waste materials should not be allowed to accumulate at the place they are created; instead, they must be cleared up and deposited in skips at regular intervals. Increased fire risk could be another consequence of allowing waste to accumulate.

The Construction (Design and Management) Regulations place a legal duty on the employer, the self-employed and any other person who controls the way in which construction work is carried out (which in this context includes the management of waste).

For further information refer to GE 700 *Construction site safety,* **Chapter E03 Waste management.**